THE ALCHEMY OF AIR

ALSO BY THOMAS HAGER

The Demon Under the Microscope:
From Battlefield Hospitals to Nazi Labs, One Doctor's Heroic
Search for the World's First Miracle Drug

Force of Nature: The Life of Linus Pauling

The

ALCHEMY

of

AIR

A Jewish Genius, a Doomed Tycoon, and the Scientific Discovery
That Fed the World but Fueled the Rise of Hitler

THOMAS HAGER

HARMONY BOOKS
NEW YORK

HARMONY BOOKS is a registered trademark and the Harmony Books
colophon is a trademark of Random House, Inc.

Library of Congress Cataloging-in-Publication Data
Hager, Thomas.
    The alchemy of air: a Jewish genius, a doomed tycoon, and the scientific
discovery that fed the world but fueled the rise of Hitler / Thomas Hager.
    Includes bibliographical references and index.
    1. Haber, Fritz, 1868–1934. 2. Bosch, Carl, 1874–1940. 3. Chemists—
Germany—Biography. 4. Nitrogen fertilizers—History—20th century.
5. Technological innovations—History—20th century. I. Title.
    QD21.H26 2008
    540.92'244—dc22
    [B]   2008003192

ISBN 978-0-307-35178-4

Printed in the United States of America

Design by Lauren Dong

10 9 8 7 6 5 4 3 2 1

First Edition

WITHDRAWN

FOR LAUREN

Ye instruments, ye surely jeer at me,

With handle, wheel and cogs and cylinder.

I stood beside the gate, ye were to be the key.

True, intricate your ward, but no bolts do ye stir.

Inscrutable upon a sunlit day,

Her veil will Nature never let you steal,

And what she will not to your mind reveal,

You will not wrest from her with levers and with screws.

—JOHANN WOLFGANG VON GOETHE, *Faust*

Contents

Introduction: Creatures of the Air   *xi*

I. THE ENDS OF THE EARTH   *1*

II. THE PHILOSOPHER'S STONE   *63*

III. SYN   *149*

Source Notes   *283*
Bibliography   *293*
Acknowledgments   *305*
Index   *307*

Introduction:

Creatures of the Air

THIS IS THE STORY of two men who invented a way to turn air into bread, built factories the size of small cities, made enormous fortunes, helped engineer the deaths of millions of people, and saved the lives of billions more.

Their work stands, I believe, as the most important discovery ever made. See if you can think of another that ranks with it in terms of life-and-death importance for the largest number of people. Put simply, the discovery described in this book is keeping alive nearly half the people on earth.

Most people do not know the names of either the men or their invention. But we should thank them every time we take a bite of food. Their work lives today in the form of giant factories, usually located in remote areas, that drink rivers of water, inhale oceans of air, and burn about 1 percent of all the earth's energy. If all the machines these men invented were shut down today, more than two billion people would starve to death.

The reason is this: We are creatures of the air. The stuff of our bodies—the atoms that make up our skin and bones, blood, brain, and everything else—comes primarily from the atmosphere. Sometimes the route is direct, sometimes indirect.

Carbon, for instance, comes from carbon dioxide, which plants take in and turn into food. We breathe atmospheric oxygen into our blood, and hydrogen reaches us through water (along with oxygen), a substance that cycles ceaselessly among gas, liquid, and solid, evaporating up into the clouds and precipitating down into our mouths. These three elements, carbon, oxygen, and hydrogen, constitute more than 90 percent of our bodies by weight. You could say we are air made solid.

But the most important element in many ways for humans is the fourth most common in our bodies—and the hardest to find in nature (at least in forms we can use): nitrogen. You can't live without nitrogen. It is stitched into every gene in your DNA and is built into every protein. If you don't get enough nitrogen, you die. It is not only necessary but more interesting than the other major elements. Chemically, nitrogen is something of a trickster, a bit promiscuous, eager to join with many other types of atoms with various types of bonds, in many different ways. Nitrogen gives proteins their twists and turns, and it confers on living molecules much of their individuality and flexibility. Nitrogen spices up the party.

The absolute necessity of nitrogen for life leads to a paradox: We are swimming in nitrogen, but we can never get enough. Nitrogen gas makes up almost 80 percent of the earth's atmosphere. We breathe it in and breathe it out all day long. But none of this huge store of atmospheric nitrogen—not a single atom of it—can nourish any plant or animal. It is inert, unavailable, dead. Plants and animals, including humans, require nitrogen in a different form, a form scientists call fixed nitrogen. The availability, or more commonly lack, of fixed nitrogen is so important that it serves as a cap on life on earth, a "limiting factor" for plant systems (and hence for animals as well, since all animals depend in one way or another on plant eaters). Put in simple agricultural terms, if you put more fixed nitrogen on a field, you can grow more. Farmers have long been nitrogen experts, feeding their fields with rotting plants (compost rich with fixed nitrogen) and animal dung (manures rich with fixed nitrogen), and rotating their crops, growing peas and beans every few years because these plants carry bacteria on their roots that make fixed nitrogen available. The secret of successful farming is moving nitrogen around.

Over our heads is a vast vault of unusable nitrogen; under our feet, a limited amount of fixed nitrogen. Nature offers only two ways of fixing nitrogen, getting it out of the air and into living systems: those special bacteria on the roots of peas, beans, and a few other plants, and bolts of lightning. Both methods produce small amounts of fixed nitrogen that accumulate slowly. As a result, humans have always been faced with a shortage of usable nitrogen—like people floating on a vast sea, dying of thirst.

The problem has grown along with the human population. Thanks to these extraordinary machines, we are doubling the amount of nitrogen available to living systems. This fundamental change has made it possible to feed billions more people than the earth could support otherwise. But we are also subjecting our planet to a huge experiment—flooding it with nitrogen taken from the air—poisoning rivers and lakes, killing swaths of the ocean, and boosting global warming, without any clear picture of what we're doing or how it will turn out. All of this can be traced back to two relatively unknown men and their machine.

But the most direct measure of their importance can be read in the lines on graphs showing the increase in human population over the past hundred years. At the turn of the twentieth century, the world supported about one billion people. Today we're above six billion and counting. If we all ate simple vegetarian diets and farmed every acre of arable land as wisely as possible using the best techniques of the late 1800s, the earth could support a population of around four billion people. In theory, the other two-billion-plus inhabitants should be starving, the natural result of population outstripping food supply, as doomsayers from Thomas Malthus to Paul Ehrlich have long predicted. But despite the extra mouths, we are not hungry. In fact the average human today is eating better than he or she did a hundred years ago, with diets that are both more varied and higher in calories. That is true everywhere, not just in the United States. (Yes, pockets of starvation still happen, but it is not because of a lack of food. There is plenty of food around. People starve because of problems shipping it to the places it is needed.) Instead of facing worldwide famine, we are dealing with a global epidemic of obesity.

The reason is the Haber-Bosch system. Haber-Bosch plants are why food today is so plentiful and so relatively cheap. Haber-Bosch machines grow the plants that feed the animals and produce the oils, sugars, meats, and grains that are making us all fat. If you want to know why so many people are putting on so many pounds today, you know where to look.

Food, however, is only part of the story. Remember the terrorist bombing of a federal building in Oklahoma? The explosive used was a couple of tons of nitrogen fertilizer boosted with another nitrogen-containing compound. Fertilizer and explosives are very close in structure—so close that one can often be used for the other. With a little chemical tinkering, the fertilizer from Haber-Bosch factories was turned into gunpowder and TNT. That meant that the same discovery that could feed the world could also destroy it. Haber-Bosch technology was used to make the explosives that killed millions in both world wars. Without Haber-Bosch, historians say, Germany would have run out of arms and surrendered two years earlier than it did in World War I. Without Haber-Bosch, Hitler would never have been much of a threat.

That was just the start. The Haber-Bosch technology was also used to make synthetic fuels. Decades before today's energy crisis, Bosch's factories were fueling Germany with synthetic gasoline made from coal. Hitler depended on it in World War II, using Bosch's synthetic fuels to gas and lubricate his planes and trucks, while at the same time using Haber-Bosch synthetic nitrogen to make his bombs and gunpowder. The story of synthetic gasoline, including prewar deals among IG Farben, Standard Oil, and the Ford Motor Company, is also told in this book.

I RAN ACROSS this story while researching a previous book, *The Demon Under the Microscope*, about the discovery of the first antibiotic. That discovery was made in the Bayer laboratories in Germany, which later became part of an infamous cartel called IG Farben, which was the world's largest chemical company until it was broken up after World War II. Farben was used by the Nazis to produce everything from rubber for Wehrmacht tires to gasoline for Luftwaffe fighters. Farben powered Hitler's mad dreams.

Farben's first director, Carl Bosch, led me into the story. I learned quickly that he was a man of contradictions: a business mogul who won a Nobel Prize and an ardent anti-Nazi who founded and led a most infamous Nazi firm. I also discovered that Bosch is one of the great mystery men of the twentieth century. He kept a very low public profile, avoided meetings, seemed to prefer machines to people, and hid away as much as possible in his Heidelberg villa, which he turned into a private playground equipped with personal laboratories, museum-quality collections, and a research-grade astronomical observatory. He hated what Hitler did to his nation, burned his private papers, died a broken man, and was forgotten.

As I learned what there was to learn about Bosch, I also started reading about the second man who developed the machine, Bosch's co-discoverer (and fellow Nobelist for his work on nitrogen) Fritz Haber. Haber was as public as Bosch was private, a scientist who reveled in attention, sought glory, drank, smoked, partied, hobnobbed with royalty, and loved making an impression in his hand-tailored military uniform. He was also Jewish. I found much more about Haber, the subject of several biographies and even a character in a couple of plays, but he too had his mysteries. How could a man who helped feed the world also be attacked as a war criminal after World War I? What was he doing in a secret laboratory hidden on an ocean liner? Was it really Haber who developed the poison gas used at Auschwitz?

These men were giants of science. Their careers took off after their invention of the air-to-bread machine and they both attempted even bigger things. They pioneered new ways of doing science, built city-sized factories, controlled world markets, and made life-or-death decisions. In addition to everything else, Haber and Bosch are in many ways responsible for creating the modern chemical industry. Their work is still vital today, not only because we live on the food they made possible but because we are just beginning to come to grips with the ecological impact of their discoveries.

FROM THE BEGINNING, from the time when humans first tamed fire to the time they started growing grain, from the wings of Icarus to the artificial heart, humans have sought endlessly to cross nature's boundaries, to break its limits, to make themselves more comfortable, more healthy, more powerful than nature alone could. This interplay between human aspiration and natural bounds has its own literature (from the myth of Prometheus and Mary Shelley's *Frankenstein* to superhero comic books and mad-doctor movies) and its own field of play. I would call that field of play—the place in which humans test their natural limits and often break them—science.

Stories about scientists are often paeans to selfless men and women bettering mankind's lot in a ceaseless march of progress. Those elements are real and are part of this story. But I have tried to create a different kind of book, one that shows what happens when scientific altruism comes head to head with politics, power, pride, money, and personal desire. This, to me, is the real world of science.

I

The
ENDS
of the
EARTH

CHAPTER I

THE PROPHECY WAS made in the fall of 1898, in a music hall in Bristol, England, by a thin man with a graying, neatly trimmed beard and a mustache waxed to alarmingly long, needlelike points. His audience, the cream of British science, thousands of formally dressed men and bejeweled women, were seated in a low-rent venue, what Americans would have called a vaudeville palace—a last-minute substitute for an academic auditorium that had burned down—but they dutifully filed in and filled every seat from the orchestra pit to the highest balcony. The hall was uncomfortably hot, especially in the upper seats. Exquisitely gowned women began opening their fans. Evening-coated men began murmuring to their neighbors that it looked as if it were going to be a long evening.

The speaker was Sir William Crookes, 1898's incoming president of the British Academy of Sciences. Impeccably dressed, erect and resolute, he looked every inch the triumphant, newly knighted physicist he was: inventor of the Crookes Tube (a predecessor of the cathode ray tubes used later for televisions and computers), recent discoverer of an interesting new addition to the periodic table that he had named thallium, fearless explorer of science, even out to its furthest edges—Crookes was an active researcher in the area of séances and the question of life after death.

Inaugural speeches were often deadly dull. The incoming

presidents of scientific associations almost always droned long lists of achievements made during the past year, with nods to numerous individual researchers, sprinkled with homilies about the importance of science for the British Empire. Crookes, however, had decided to shake things up. He adjusted his oval glasses, glanced at his notes, looked up, and got right to the point. "England and all civilized nations," he said, "stand in deadly peril."

The fans in the balcony stopped fluttering. Crookes's voice was clear but he spoke softly. The hall went silent, the audience straining to hear as the speaker continued. If nothing was done soon, he explained, great numbers of people, especially in the world's most advanced nations, were soon going to begin starving to death. This was a conclusion that he was forced to accept, he said, after considering two simple facts: "As mouths multiply," he said, "food sources dwindle." The number of mouths had been increasing for some time thanks to advances in sanitation and medical care, from the installation of improved water systems to the introduction of antiseptics. These were great triumphs for humanity. But they carried with them a threat. While population increased, land was limited; there were only so many farmable acres on earth. When every one of those acres was under the plow and farmed as well as it could be, the population would keep going up, the farmed and refarmed soil would slowly lose its fertility, and mass starvation would, of necessity, ensue. His research led him to estimate, he said, that humans would begin dying of hunger in large numbers some time around the 1930s.

There was only one way to stop it, he said. And then he told them what it was.

EVERY AGRICULTURAL SOCIETY in every age has had its own methods, rites, and prayers for ensuring rich crops. Homer sang of farmers gathering heaps of mule and cow dung. The Romans worshipped a god of manure, Stercutius. Rome made an early science of agriculture, ranking various animal excrements (including human), composts, blood, and ashes according to their fertilizing power. Pigeon dung, they found, was the best overall for growing crops, and cattle dung was clearly better than horse manure. Fresh human urine was best for young plants, aged urine for fruit trees.

Both the Romans and the ancient Chinese also understood that there was another key to a healthy farm: crop rotation. No one knew why or how it worked, but never planting the same crop twice consecutively in the same land, instead alternating it with certain crops like peas and clovers, managed to replenish the fertility of fields. Every few years the Chinese made sure to rotate in a crop of soybeans; chickpeas were the crop of choice in the Middle East, lentils in India, and mung beans in Southeast Asia; and Europeans used peas or beans or clover. "Oats, peas, beans, and barley grow" was more than a children's rhyme. It was a timetable for successful farming.

Healthy farms had compost pits, plenty of domestic animals for manure, and a system of crop rotation. But it was never enough. It took scores of tons of manure per acre to grow great crops. Manure gathering and handling grew into a small industry, employing thousands of workers who scoured the countryside for cow and pig excrement, cleared city streets of horse manure, and then sold it by the stinking ton to farmers and gardeners. There was never enough. A heavy application of manure helped for a season or two, but then the fertility of the soil declined and more was needed. In the most intensively cultivated land in Europe—the Marais district of Paris—owners of small city-garden plots applied dung at rates as high as hundreds of tons per acre, and every year they had to repeat the process. By 1700 or so, hungry Europeans were experimenting with other soil additives in an attempt to increase their yields, trying sea salt, powdered limestone, burned bones, rotting fish, anything that might keep their soils producing.

But the world's best farmers were not in Europe. In the wet, warm farmlands of southeastern China, farmers a millennium ago were already expert in using every possible kind of fertilizer, hoarding their human waste and adding it to the output from their domestic animals, composting vegetable scraps and leaves, and tossing in seed cakes to enrich their fields. It was all applied to the most ingenious farm system imaginable: a complex of dike-and-pond fields in which they grew not only rice, mulberries, sugarcane, and fruits but also carp. The fish waste helped fertilize the crops. The dung of the water buffaloes used to work the fields helped fertilize the crops. So did the waste of the ducks that swam in the ponds. They grew a native water fern in the paddies that acted like a crop of soybeans, adding fertility to the soil. The tropical climate allowed multiple harvests per year. This was the highest-yield traditional agricultural system ever devised. Using it, the Chinese could feed as many as ten people with the output from each acre of farmland, a yield of food five to ten times higher than the European average of the 1800s. "The Chinese are the most admirable gardeners," an appreciative European scientist wrote in 1840. "The agriculture of their country is the most perfect in the world."

IT WAS NOT enough. During the nineteenth century, millions of people left the farm and flocked to cities during the Industrial Revolution. As the cities grew and the population of the earth rose faster and faster, it became clear that feeding ten people per acre, the pinnacle of traditional agriculture, was nowhere near good enough. The crisis Crookes predicted would have happened fifty years before his speech, but for the opening of vast new farming territories, from the Great Plains of the United States and the steppes of Russia to the vast landscapes of Australia. When their land played out, farmers simply moved west or south or east to the next expanse of virgin soil.

Now, however, Crookes warned, the earth held no more Great Plains. The globe had been explored, mapped, and the best agricultural areas settled and plowed. From this point on, farmers would have to make do with the land they had, refarming the same acres year after year. This brought Crookes to the critical issue: When land was farmed repeatedly, no matter how carefully crops were rotated, no matter how scrupulously every bit of animal dung was applied, the soil slowly lost its original fertility.

His analysis focused on wheat, the staple of Europeans and North Americans, the staff of life for Caucasians. Any drop in wheat production threatened, as he put it, "racial starvation." His conclusion, based on what he called stubborn facts, seemed incontrovertible: In a few decades, the populations of the great wheat-eating peoples—including the Caucasians of the British Empire, northern Europe, and the United States—would outstrip their grain of choice, and thousands of people, then hundreds of thousands, then millions, would begin to die.

The best traditional farming techniques in the world were not enough to avert the coming crisis. England itself was using the most advanced farming techniques, the best possible mix of crop rotation, animal manuring, and composting, and the English, he said, would be starving now if they did not import tons of grain from other nations. What would happen when those other nations, in order to feed their own growing populations, stopped exporting?

There was only one answer, Crookes said: the creation of vast amounts of fertilizer, new fertilizer by the thousands of tons. As there was not enough natural fertilizer in the world to meet the needs of the coming twentieth century, some way would have to be found to make more, to make it synthetically, to make it in factories. Finding new ways to make fertilizer, discovering and making what he called chemical manures, Crookes told his audience, was the great challenge of their time.

"It is through the laboratory," he said, "that starvation may ultimately be turned into plenty." He then pinpointed the kind of scientist who would save humanity. "It is the chemist," he said, "who must come to the rescue.... Before we are in the grip of actual dearth the chemist will step in and postpone the day of famine to so distant a period that we and our sons and grandsons may legitimately live without undue solicitude for the future."

EMPTY A BAG of store-bought fertilizer and what pours out is usually a mix of three elements, N, P, and K—nitrogen, phosphorus, and potassium—the three most essential nutrients for plants. None of today's major crops can survive without them. All of them can be found, in varying but generally low amounts, in manures and composts. It is because of these nutrients that compost and dung were the farmers' best friends.

For most crops, the most important of the three is nitrogen. Atoms of nitrogen are stitched into every protein and every bit of DNA and RNA in every cell of every plant (and every animal). Without nitrogen, life is not possible. This single element is so important that its availability represents a limiting factor for most plant ecosystems, which means that the availability of nitrogen pretty much determines how much will grow. Low nitrogen equals low yields. High nitrogen equals big crops. It seems just about that simple.

But it's not. Plants need nitrogen, but they are picky about the kind of nitrogen they use. Almost 80 percent of the air around us is nitrogen, for instance—we are swimming in an ocean of it—but it is locked up, unavailable to plants and animals. They have no mechanism for absorbing and metabolizing atmospheric nitrogen.

Plants use fixed nitrogen, nitrogen in a chemical form different from how it exists in the air, usually a solid or liquid form.

Manure is a source of fixed nitrogen, and so is compost—which is why they make good fertilizers. Uncultivated, virgin soils have fixed nitrogen stored in them; this is why pioneers' first few crops in a new land are generally the best they'll ever see. As the crops use up the fixed nitrogen, the amount in the soil drops, and so does fertility. Crops become sparser, plants punier, and yields lower.

"Wheat preeminently demands nitrogen," Crookes told his audience. But centuries of wheat farming had depleted the soil's stock of fixed nitrogen, and farmers were unable to adequately replace it. Using current farming practices, he warned, "We are drawing on the earth's capital, and our drafts will not be honored perpetually."

CROOKES KNEW THAT his remarks might come as something of a surprise to his audience. Most Englishmen believed that there was in fact no fertilizer shortage at all. Fertilizer was available in plenty; it came in canvas bags, delivered by the tens of tons from South America, shiploads of fertilizer unloaded at docks in all English ports. British farmers had been swearing by it for decades. First, in the 1840s, there had been mountains of South American bird guano, which many European farmers were convinced was the best fertilizer in the world, then later Chilean nitrate, a very clean, white fertilizer, mined somehow from the desert wastes somewhere near the Andes. It was magical stuff, the nitrate, excellent fertilizer, granular, easy to apply, raised yields enormously. Dizzying fortunes had been made trading nitrate stocks in London. South America was full of fertilizer, was it not?

Crookes carefully explained that indeed there was an end to the South American supply, and it was coming soon. Wheat growers had become increasingly dependent on the Chilean product, spreading it on their fields by the hundreds of thousands of tons per year. Such use simply could not be sustained. He ran through more numbers, showing that if current trends continued, the Chilean nitrate fields would be exhausted within decades, perhaps by the 1920s, certainly by 1940. When that happened, the game was up. With no more big sources of fertilizer, yields would plummet and people would starve—unless scientists could come up with an answer.

He ended by calling his fellow researchers to action. The only answer, he said, was to find a way to make synthetic fertilizers—fixed nitrogen—refining it from the earth's greatest reservoir of nitrogen: the atmosphere. Other scientific discoveries might make life easier, might help build wealth, might add luxury or convenience to the lives of the wheat-eating peoples, but the necessary discovery, the vital discovery—the discovery of a way to fix atmospheric nitrogen—was a matter of life and death. "Unless we can class it among the certainties to come," he said, "the great Caucasian race will cease to be foremost in the world, and will be squeezed out of existence by races to which wheaten bread is not the staff of life."

Crookes's racism was as naked as it was common. In 1898 most Englishmen took it for granted that they represented the pinnacle of civilization. His audience was English and he spoke to their native prejudice, using their chauvinism as another way to drive home his point. In fact, the same "stubborn facts" applied to other races as well. The entire population of the world, whether it ate wheat, rice, corn, or millet, needed fixed nitrogen. Whoever found a way to create it out of the air would not only save humanity but would likely become very, very rich.

"A brilliant success," Crookes wrote a friend a few days after his speech. "I am overwhelmed with compliments, several old stagers saying it was the best address that they had ever heard." He did not mention that a fair portion of those in the upper balcony exited about halfway through his eighty-minute talk, fleeing the stifling heat. Those who stayed, however, including some reporters, were impressed. Word of mouth turned his presentation into a sensation. News of the impending doom of the Caucasian race rippled out from Bristol through England, then to newspapers around the world. His words were read not only by scientists but economists, politicians, intellectuals, and businessmen. Some experts chimed in supportively; others were critical. It was very much like today's global-warming debate. The publicity all helped boost Crookes's presentation into the ranks of the most influential public addresses of the day. He received so much attention for it that he later expanded his remarks into a popular book.

Just as with global warming, the issue became a matter of controversy between those who accepted Crookes's figures and those who believed that his "crisis" was overdramatized. A few critics challenged his numbers, especially his estimates of the amount of nitrates left in Chile. They said that the South American desert was not about to run dry, that it was practically an endless sea of fertilizer.

THEY WERE WRONG. In the hundred-plus years since Crookes gave his speech, the human population has doubled, and doubled again, and will double again in the next few decades. As a species we long ago passed the natural ability of the planet to support us with food. Even using the best organic farming practices available, even cutting back our diets to minimal, vegetarian levels, only about four billion of us could live on what the earth and traditional farming supply. Yet we now number more than six billion, and growing, and around the world we are eating more calories on average than people did in Crookes's day.

Between those two points—between Crookes's dire warning and today's global epidemic of obesity—a discovery of unprecedented importance saved the world from starving. This is the story of that discovery.

Chapter 2

T HERE HAVE ALWAYS been researchers. They were called other things in ancient days—sorcerers, healers, priests, witches, shamans, and sages—but they shared the same basic drives as modern-day scientists: They all wanted to understand how the world works and to use that knowledge to make human lives easier, safer, and more rewarding. The talented men and women who discovered how to make bread and how to build the ovens it baked in were doing a primitive sort of science; and so were those who coaxed the essences of plants from leaves and seeds and used it to cure ailments; and so were those who made the first wine, smelted the first metals, crafted the first bows and arrows, pressed paper, invented mathematics, preserved food, irrigated fields, tracked the stars, domesticated animals, and dyed cloth. All these breakthroughs and many more were made millennia before we had anything like what we now call science (a word that did not exist in its modern sense until the eighteenth century).

"Alchemists" was a name applied to another group of early researchers. Alchemy's roots go back to ancient Egypt, to the search for life after death and the study of mummification. The practice evolved over time, but in a broad sense it mixed primitive chemistry and observations made from metalworking with religious and philosophical ideas, all packaged in the form of a mystic quest. The goal of alchemy was to distill the pure soul of nature from the rough, chaotic stuff of the earth, to find the spirit hidden in matter. God was in everything, even the rocks, and alchemists believed their work was a way to understand the workings of the divine. They discovered that wonderful things happened when they heated certain substances with others: sulfur, mercury, lead, salts, acids, plant extracts, metals, and minerals. Transformations took place; things changed before their eyes, gave off strange odors, and took on new colors and forms. Vapors given off by heated liquids could be collected, condensed, studied, and mixed with other substances, coagulated, dissolved, and repurified. At each step the alchemists believed they were drawing closer to the spirit of the thing. (The idea is still written into our language—the vapors given off by heated mashes of fermented fruits and grains, for instance, are condensed into what we still call spirits.) Their goal was to approach pure perfection on earth, to create in their primitive laboratories a divine substance capable of curing ills, conferring long life, and transforming base elements into gold. In the West they called this the Philosopher's Stone.

A similar search went on in the East, too, where Asian alchemists also sought an elixir of immortality, a mix of substances that could stop or reverse the aging process. In China, more than a thousand years ago, an alchemist engrossed in the search for immortality stumbled across something that proved quite the opposite. While trying various combinations of substances, he ground up some sulfur, moistened it with honey, and then added a good portion of a white crystalline powder scraped from stone walls, most commonly those near garbage pits, barnyards, and graveyards. It was called a salt, although the word, which we now use to refer to table salt (sodium chloride), had a broader meaning in the old days (and still has a broader meaning for chemists). To the old alchemists, a salt was any sort of whitish crystal that had "sapour" or a salty taste; the Chinese salt found on walls and used in this case was potassium nitrate. The ancients knew that different salts from different places had different properties and had worked them into a number of recipes for everything from preserving meats to making medicines.

The Chinese alchemist heated his mixture and got a surprise: It blew up, shot out flames, and disappeared into smoke. News of his discovery quickly spread, and the formula for what the Chinese called *huo yao*, or "fire drug," was refined and perfected by other alchemists. Honey was soon replaced with ground charcoal. It was fascinating to work with, and dangerous. As early as the ninth century, a Taoist text noted that the alchemists' "hands and faces have been burnt, and even the whole house where they were working burned down." But properly controlled, the fire drug was a powerful and useful tool. By mixing other ground minerals and metals into *huo yao*, Chinese alchemists found they could make flames of different colors, from deep blue to blinding white, making possible the first fireworks. The mixture could be put into a bamboo tube and attached to the end of a lance, then lit and thrust into the faces of one's enemies, making a terrifying weapon. It could be packed into paper tubes and lighted with a wick—the first firecrackers. Rammed into metal cylinders with a ball on top, it could hurl the ball across a field—the first cannons and muskets. Always, however, the most important ingredient, making up three-quarters of the fire drug, was the white salt. Without it, no fire drug could be made.

By the late Middle Ages, the recipe had spread across the Middle East to Europe. There they found the same salt—which they called China Snow or nitre—growing on stone walls, especially underground, in basements and crypts. The Romans called it *sal petrae* (Latin for "salt of stone"). That led to the most common name for the important salt: saltpeter. The Chinese *huo yao* became the West's gunpowder.

Gunpowder was the atomic bomb of its day. It changed the

nature of war. Muskets loaded with it could destroy armies of armored knights; cannons using it could batter stone castles into rubble. Gunpowder marked the end of the feudal age. By the 1500s Europe was embroiled in an arms race, with every nation seeking ways to make more and more gunpowder. Because every barrel of gunpowder required three-quarters of a barrel of saltpeter, access to saltpeter became a matter of national survival.

The problem was that there was not a lot of it. Once scraped from stones and earth, it grew back, but with frustrating slowness. It took years to build up enough to warrant gathering. In England, by royal decree, the king's saltpeter gatherers—Petermen, they were called—scoured the countryside for the white salt. They quickly became among the nation's most reviled and feared public officials. Wherever they found saltpeter, the Petermen were warranted to dig it out, regardless of whose property it was on, regardless of the difficulties involved, even if it meant moving a privy, tearing apart a stable, or ripping up the floor of a house. Petermen opened pits, tore down walls, and commandeered carts and horses. They often took bribes.

They could never find enough to please the Crown. Naturally occurring saltpeter formed too slowly, too haphazardly, to supply a nation's needs. So ways were found to speed its growth. By the eighteenth century the art and craft of the "saltpeter plantation" had been perfected, a way to hurry nature by creating an artificial environment for the rapid formation of saltpeter: a series of shallow trenches lined with clay, filled with a mix of soil, manure, garbage, and ashes, heaped into something that looked like a long burial mound, and moistened with sewage and urine. After months of ripening in the sun, a crust of saltpeter would begin grow out of the earth in the form of fine crystals that looked like tiny white flowers. A well-run plantation could produce, every two years, five or ten pounds of pure saltpeter per cubic meter of dirt. It was painstaking, dirty, labor-intensive work. But it was necessary for national survival. In 1626 England's King Charles I commanded his subjects to "carefully and constantly keep and preserve in some convenient vessels or receptacles fit for the purpose, all the urine of man during the whole year, and all the stale of beasts which they can save" to be donated to the saltpeter plantations. In Massachusetts Bay Colony, an act required every large farm to erect a nitre shed. In Sweden, farmers paid a portion of their taxes in saltpeter.

It was still not enough. A few large natural deposits were found in caves, and the world was searched for more. Significant amounts were found in the rocky terrain of Spain, Italy, Egypt, and Iran. The mother lode of saltpeter, however, the only natural deposits in the world large enough to feed the gunpowder needs of an entire nation, was discovered in the mud flats of the Ganges in India (where it was believed that a combination of the river water, the hot climate, and the dung of holy cows combined to create a sort of huge natural saltpeter plantation). The British East India Company started shipping it to England by the ton in the mid-seventeenth century—it was one of the company's most important cargoes—and this vital natural resource made India an especially important target for European colonial expansion. Saltpeter was a significant factor in favor of the British takeover of India.

WHAT DOES GUNPOWDER have to do with growing things? The answer came from South America, and the story goes like this: A small party of Indians, during the time of the Spanish occupation in the 1700s, was trekking from the Andes Mountains down to the villages on the Pacific to do some trading. To reach the coast, they had to cross a wide strip of desert called the Tarapacá. Here, in the silent, eerie wastelands, they camped for the night, building a fire from dried cactus to keep warm. To their astonishment, the earth itself caught fire, with flames and sparks shooting up from the ground around the fire. They thought it was the work of the devil. They ran. When they returned the next morning to gather their things, one of the group, a wood-cutter named Negreiros, gathered some of the earth from the area of the fire. This he dutifully brought to a nearby mission and gave to a priest, along with a description of the events of the night.

The priest knew enough about chemistry to ascertain that the chunks of earth were rich with a kind of salt, a mineral similar to the saltpeter used in gunpowder, although not as powerful. It was a sort of inferior saltpeter, not the "true" saltpeter, the sort found in India and China, used for gunpowder. Still, it was capable of catching fire. He assured the travelers that the devil had nothing to do with their experience. Then he threw the rest of the sample out. Some weeks later the priest noticed that the vegetation growing around the spot where he had thrown the devil's dust was greener, taller, and more luxuriant than the other mission plantings. He told this interesting fact to a visiting British naval officer. At this point two things happened, according to the legend. One was that a small local industry grew up to refine the desert saltpeter to be sold for use in fireworks and gunpowder. It was not a perfect substitute for China Snow— true saltpeter was rare in South America—but the Spanish found that the inferior local salt was at least good enough to make a cannon fire. The other was the beginning of the idea that a salt from the desert might have use as a fertilizer.

No one knows how much of the legend is true. What is known through historical records is that by the early nineteenth century at least two scientists had visited the area of the Tarapacá and written of the properties of the local saltpeter, which the natives called *salitre*. The French had started importing some to make gunpowder. Interest in the South American salt-peter was lower in Britain, where the Crown controlled the world's richest stocks of high-quality true saltpeter in India, but a few bagfuls of the South American variety had reached the area around Glasgow, where farmers found that adding it to their fields increased their crop yields.

ENOUGH WAS KNOWN about it to vaguely interest a young British divinity student in the summer of 1835. His name was Charles Darwin. He was standing on the deck of the small surveying brig HMS *Beagle*, riding the swells at anchor in one of the few sizable bays to be found on the bleak thousand miles of coastline that stretched up the west coast of South America from its southern tip to the Spanish holdings in Peru. He was eager for something, anything new. Week after week as the ship had rounded the Horn and sailed slowly northward he had seen little but dry, brown hills. Now, after endless days of tacking, he was completely bored by the repetitious, often vegetationless coastline. "The coast was here formed by a great steep wall of rock about 2,000 feet high," Darwin wrote dutifully in his notebook. The rocky hills for the most part rose directly from the sea, leaving little room for inhabitants—not that the area could support many humans anyway. The sea was beautiful here and full of life, but the land was as arid and uninviting a stretch of desolation as Darwin had ever seen. A few scattered Indian tribes survived by fishing, but the hills for hundreds of miles were so rugged, the flat areas near the water so rare, the sources of fresh water so few that he felt he was looking at a near-perfect desert in both senses of the word: dry and empty. The *Beagle* was anchored in the bay of one of these small fishing villages, the only one for leagues with a harbor of any size, a place called Iquique. He was ready to stretch his legs. He had heard stories about the interior, about a strange local industry practiced on the other side of the hills, and he wanted to see it in person.

A voyage around the world in 1835 generally took two to

three years. Darwin's time on the *Beagle* would stretch to five. Five years with a few dozen uneducated companions in a small floating box, facing months of tedium punctuated by moments of terror, could sometimes drive officers mad. A few years earlier, the captain of a previous voyage of the *Beagle* had committed suicide in the wastes of Patagonia. The Royal Navy did not want that to happen again. Darwin—young, energetic, well schooled, interested in the nature of bugs and plants as well as the nature of God—had been taken on board as an unpaid gentleman's companion. He could take as many scientific notes as he liked, but his main job was to amuse the captain.

Darwin's Iquique notes give a sense of his mood after weeks of routine and tedium. "The aspect of the place was most gloomy," Darwin wrote, "the little port with its few vessels and a small group of wretched houses seemed overwhelmed and out of all proportion with the rest of the scene. The inhabitants are like those on board a ship, everything comes from a distance." The village was nominally in Peru, not that it mattered much. The area had been ruled for thousands of years by scattered tribes, then for a time by the Inca, then by the Spanish, now theoretically by the new republican government of Peru, but the reality was that Iquique was so remote, so poor, so undesirable, that no one seemed to care much who governed it. To the west was the sea, to the east the high Andes. Far to the north was Lima, the great city of the Spaniards. To the south was a desert region, the Atacama, claimed by Bolivia. Somewhere beyond that, far to the south, lay another desert claimed by another of the new republics, Chile. The people of Iquique referred to their own stretch of South America by its old Indian name, Tarapacá.

What mattered in the Tarapacá in 1835 was not politics but survival. It rained in Iquique about once every twenty years. No crops could grow. Every drop of drinking water had to be shipped from a spring forty miles away; villagers in Iquique bought it by the bottle or, when they could afford it, by the barrel. They survived only because their village perched on the edge of one of the world's richest fishing grounds. The great Humboldt Current, a giant river in the sea, rolled from Antarctica up along the west coast of South America, bringing with it a wealth of plankton, shrimp, and fish that fed local seals, natives, and millions of sea birds. The land here was dead. But the sea was teeming.

On July 13, the day after the *Beagle* anchored, Darwin hired an Iquique guide to take him on a mule trip over the wall of rock and into the interior. It took them most of the first day to zigzag up a dusty trail to the summit. "Very tedious," Darwin wrote. The sun was setting by the time they topped the hills and he saw what lay beyond: a wide plain of broken rock and low reddish hills spreading east all the way to the Andes, purple and white in the distance. There was not a tree in sight. There was not a blade of grass. The only sound was the sighing wind. This lifeless expanse, as barren as the surface of Mars, was the real Tarapacá.

When most people think of deserts, they see the sandy dunes of the Sahara, scenes from *Lawrence of Arabia*, or perhaps the sagebrush and sandstone of a John Wayne western. The Tarapacá is nothing like that. It is a blasted land, scarred and wasted, like something out of an uncomfortable dream. There is no sagebrush. There are no plants at all. As they rode through the dusk Darwin could see bones and the skins of dead mules along the side of the trail, attended by a few vultures, the only local wildlife he would see on this trip. "This is the first true desert [*sic*] I have ever seen," Darwin wrote. "The effect on me was not impressive." His attention was caught, however, by what looked incongruously like the remains of thawing snow, whitish patches scattered here and there across the desert. It was not snow, of course—snow would mean water, and there was no water—but a thick layer of some whitish mineral. It had a salty taste. In some cases it lay on top of the ground; in some cases it dove under, buried under a layer of gravel and dirt. But the more Darwin looked, the more he saw. It was everywhere. "The quantity is immense," he wrote. He supposed that it might be the salty residue of a great inland sea that long ago had dried up.

He and his guide took shelter as night fell in the house of a man who ran a small factory—an *oficina*, it was called—devoted to gathering and purifying the local salt. They were welcomed, given food, and a place to sleep. The whitish crust Darwin had seen on the desert, his host explained, was the source of the local saltpeter, the *salitre*. The raw deposit of salt, the thick, whitish, rocklike crust he had seen in the desert, was called *caliche*. In some places the *caliche* was four or more feet thick. Darwin's host made his living by hiring workers—local Indians mostly—to find the best *caliche*, break it up, bag the chunks, load them onto mules, and bring them to the man's homemade factory. Here other workers crushed it, shoveled it into cow skins, and soaked it in water for a day. The water had to be packed in on muleback. Then they emptied the slurry into iron pots and boiled the contents over cactus fires. In some *oficinas* they added a beaten egg to the pot to help clean out impurities. The good part of the salt, the valuable part, Darwin's host explained, dissolved in the hot water, which was strained into clay pots to cool. The next day, they could see, a layer of crystals would form in the bottom and on the sides of the pots; this was what they wanted, the saltpeter they could sell, the *salitre*. They poured out the water, scraped out the *salitre*, and spread it on canvas to dry in the sun. It looked something like coarse table salt. This they bagged for a mule trip down to the port at Iquique.

How long this had been done, no one really knew. There were legends about it, stories of Indian shamans who hid in the hills and made it during the time of the Spanish, stories of the Inca using it, but who knew? These days the *salitre* was used for many things. Some people, it was said, made soap with it, others fireworks, and now the Europeans were beginning to use it to make things grow. The *salitre*, Darwin learned, was what chemists called sodium nitrate. The people on the docks often called it, simply, nitrate. In Darwin's day, a hundred-pound bag of nitrate, packed by mule to the docks at Iquique, sold for fourteen shillings.

He had seen enough. Darwin was more interested in living things than he was in primitive chemistry. His boredom allayed for a day, he jotted a few more notes, spent the night, and returned to the *Beagle* eager for the next leg of his journey, a visit to what promised to be a more interesting port of call, a group of islands called the Galápagos.

Chapter 3

Souterar American nitrate was a good fertilizer, but the fast-growing and increasingly hungry European nations during the Industrial Revolution fixed their attention not on the desert but to the sea a few hundred miles north of the Tarapacá, to a place where there were mountains of the most perfect and powerful natural fertilizer ever found. During the twenty years from the 1840s to the 1860s, it fueled national economies, determined foreign policy, caused a war, and earned a lasting reputation as one of the world's oddest industries.

The first thing approaching sailors saw was a pale yellow pall over the water. As their ships drew closer the islands came in sight, a few whitish dots on the horizon. Then wheeling clouds of seabirds. Closer still, if the wind was right, the sailors could smell it: the rich scent of newly turned earth, an undertone of low tide on a hot day, a bit of old outhouse and rotting fish, and a sharp tang of urine. To sailors who had been at sea for a few months, the odor might have seemed interesting, for a few minutes. After a few days anchored near the islands, it would grow unbearable. Closer still they could make out the line of shanties on the islands' snowy tops. The smell grew stronger. Then a guide boat came out to lead them to their anchorage, and they eased into a forest of masts bobbing around the rocks.

Each ship took its place in line and waited. It could take weeks, rolling on the jade green water, trying to ignore the stench, before they took their turn for loading. The only way to speed things up was to bribe the islands' "governor." The only landing spot was a ledge in front of a great sea cave, hollowed out of the hundred-foot-high cliffs. The swells rose and crashed on the rocks. Visitors had to time their leap from boat to ledge perfectly, jumping at the top of the rise. Supplies were handed into baskets and lifted to the top with winches and tackle. Visitors had to crisscross their way up systems of trails and ladders that snaked up the cliffs. At the top was the "palace," a flat-roofed wooden shanty furnished with rough cots, a few writing desks, some German maps, and a couple of old pistols hanging on the walls. A large dog panted in the yard. The boss of the islands was a red-haired, epithet-spewing Hungarian. It was to him that the ship captains came courting, one after another, bearing gifts of sardines, cold hams, and bottles of wine in a bid to buy a faster loading time.

From one side of the governor's house, wrote a visitor, the view was "enchanting. Imagine the Andes and the Pacific in one view—the islands with their precipitous walls indented with immense caves, and surrounded by fantastic rocks, fringed with foam—the pure ocean air—the myriads of sea birds—the shipping—the schools of sea lions—and almost always, far or near upon the blue waste—the spout of whales, and the white sails of ships coming or departing—altogether the scene is full of exhilaration and excitement."

Out the other side, the side with the pistols hanging near the windows, the scene was more like something from Dante's *Inferno*.

THE SLAVES WERE half dressed, most of them, with rags tied around their heads and over their mouths. Their skins were white with dust. They dug, shoveled, and pushed under the watch of armed guards who clubbed or whipped them if they slowed. The sun through the dust was blinding. There were no trees to provide shade. There was only the smell, the heat, and everywhere the screaming of birds.

These were the Chinchas Islands, a sprinkling of rocks six miles off the coast of Pisco, Peru, which constituted, in 1850, acre for acre, the most valuable real estate on earth. The value came from the ground the workers and the birds walked on: ten stories of guano, the world's best fertilizer.

The ground everywhere was springy to the step. Deep down, when the workers dug into it, it turned a rusty red. Some of the workers—about six hundred of them, all from China, coolies tricked or coerced into signing long-term labor contracts that made them virtual slaves—picked and shoveled the dusty guano into wheelbarrows. The others ran the wheelbarrows to the edge of the cliffs and dumped them into canvas chutes that funneled the guano directly into the holds of ships far below.

Visitors were appalled by what they found in the Chinchas. The coolies lived "almost naked, under the tropical sun . . . no days of rest," one observer wrote, in huts made of reed, surviving on two small meals a day, usually maize or rice and a few bananas. On good days they might get a little meat. They worked up to twenty hours a day, six days a week, in intense heat and choking dust that never settled because here, along the same coast as the Tarapacá, it almost never rained. Up to a quarter of the workforce at a time would be too sick to work, from exhaustion, exposure, malnutrition, and something called "guano handling illness," which included some combination of shortness of breath, coughing up blood, fainting, swollen legs, cramps, vomiting, and diarrhea. Scurvy was common. Sometimes a trench in the guano would cave in and a few would be buried alive. Any coolies who did not meet their quota of one hundred wheelbarrow loads a day had to make it up on Sunday, their only day of rest. If their hands were too sore and blistered to hold a shovel, they were yoked to the wheelbarrows like mules.

There was a prison ship off one of the islands where sailors loading guano could see eight or ten coolies at a time hanging on masts in the sun, without food or water, for infractions as minor as refusing an order. Others were punished by being tied to buoys where the sea washed over them all day or chained to small leaky boats and forced to bail to stay alive. An outraged British visitor wrote, "No hell has ever been conceived of by the Hebrews, the Irish, the Italian, or even the Scotch mind for appeasing the anger or satisfying the vengeance of their awful gods that can be equaled with the fierceness of the heat, the horror of the stink, and the damnation of those compelled to work there."

Until their contracts were paid off, there was only one way out. Some coolies killed themselves with opium or drink—there was a thriving drug trade in the camps. Others threw themselves off cliffs or simply swam out in the sea. Dead coolies were "buried like dogs," one visitor noted, in shallow graves scratched out of the guano. No one knows exactly how many died. Sometimes their bones were found strewn around the surface, dug up and scattered by guard dogs. For this the coolies were paid three reales (about a third of a peso) a day, two of which were withheld for meals. The contracts, held by Peruvian overseers, usually lasted five years. Guano was big money for Peru—for a few years, income from its sale supplied most of the Peruvian national budget—and few Peruvians would work on the islands, so they bought into the coolie trade, with a few wealthy families making money on contracts for the Chinese as well as the fertilizer.

Visiting ship captains and their crews hated the guano trade. At its height more than a hundred ships at a time might be waiting to load at one of the few places they could at the two largest islands. When their turn finally came, the barrow loads of guano hurtling down into their holds raised clouds of dust so foul that most of the crew escaped by climbing into the rigging, coughing. The ships, mostly from England and the United States, were covered in yellow dust. Sailors unlucky enough to draw duty as a "trimmer"—leveling the deep piles of guano in the hold—could work no more than twenty minutes at a time, even with damp cloths tied around their mouths and noses, before clambering out, gasping for breath. Some got nosebleeds. Some were temporarily blinded. The guano ship crews drank heavily, brawled, and at the height of the trade threatened so much trouble that the U.S. Pacific Fleet was ordered to patrol the area "to maintain order among the unruly merchant seamen."

THEY WERE ONCE holy islands. For a thousand years the Incas and the tribes before them had paddled from the mainland to the rocks with their living white crowns of birds to collect the earth, a powdery substance that when dug into their fields made the corn grow in abundance. They called the powerful earth *huanu*. They ranked it with gold as one of the gods' most precious gifts and made holy sanctuaries of the islands where it was found. Visiting was forbidden when the birds were breeding. Killing a seabird in the area was an offense punishable by death.

The Indians understood what they had, but it took Europeans a while to figure it out. The Spanish first heard about *huanu* (which they spelled "guano") from the sixteenth-century son of an Inca princess and a Spanish conquistador who informed Madrid that there were islands "composed almost entirely of bird guano" that from afar "appear like the peaks of some snowy mountain range." He reported, "There is much fertility in this bird-manure." His letter was ignored for two centuries as the Spanish focused their attention on looting the area's gold and silver.

Even after German naturalist and explorer Alexander von Humboldt (the man after whom the Humboldt Current is named) collected a lump of Peruvian guano and brought it back to Europe early in the nineteenth century, it took a while to appreciate what had been found. Humboldt, accompanied by a French botanist who saw amazingly fertile fields in arid Peru, heard that farmers had used guano for a thousand years, and appreciated a local proverb: "Guano, though no saint, works many miracles." Humboldt's guano was analyzed, and the chemists reported that it was extraordinarily high in urea and phosphates—a finding that seemed to excite almost no interest, scientifically or economically. When the first small commercial shipment of a few barrels of guano, shipped out by an enterprising Peruvian, arrived in England a few years later, it found no buyers and was dumped into the Thames. In 1813 the outstanding English chemist Humphrey Davy did his own analysis and reported that while guano was high in ammonia and urea, "the dung of sea birds has, I believe, never been used as a manure in this country."

That was about to change. In 1824 the editor of *American Farmer* magazine told his readers that a midshipman aboard the USS *Franklin* had arrived at port in Baltimore carrying, "among other valuable and curious things," a small amount of "that celebrated manure, *Guano dung*, possessing such astonishing fertilizing properties." Some of it made its way to the governor of Maryland, who tried it on his farm. He found it to be the most effective manure he had ever applied to corn. Word spread. In 1838 two Peruvian businessmen forwarded more guano samples to Liverpool, where "most encouraging" results were obtained among local farmers. A Scottish farmer reported that the substance boosted his yields 30 percent. Others talked of barren fields being brought back to life and trees blooming twice a year. The Earl of Derby purchased an entire shipload. A market for the stinking stuff built rapidly. Ports in England began experiencing what Southampton did when the first guano ship arrived: a stench "so foul," as one observer reported, "the entire town took to the hills."

The smell was nothing, however, compared with the demand. American plantation owners, many of whose cotton and tobacco fields were playing out, found that applications of guano could restore their fortunes. The only problem with it was that the South American guano was so powerful it had to be applied carefully to avoid "burning" crops. By the 1840s they were buying all the Peruvian product they could and started looking to other sources.

By far the best guano in the world came from the Chinchas Islands. The Peruvian government, shaky as it was (the government went through two dozen changes of regime between 1824 and 1841), began to realize what the Incas had known centuries before: Guano was as valuable as gold. The more they could sell in foreign markets, the more money would flow in to prop up whatever government was in power. The Chinchas were nationalized, with the government taking control of harvesting and shipping, handing out licenses for the coolie trade, levying taxes—and making a great deal of money. Rather than dirtying its hands developing the guano trade on its own, Peru sold contracts for the work to foreign firms in exchange for large advances on potential profits. The firms shipped coolies in and guano out. Money began to roll.

THE GUANO BOOM was on. Guano, this "best of all possible manures" became "like a necessity of life to us," wrote a British farmer, "an all but indispensable fertilizer." A U.S. expert estimated it was thirty-five times more powerful than standard barnyard manure. A man identified as one of George Washington's nephews said, "I have never seen a worn out or poor soil on which it would not, if properly applied, bring good crops." *Farmer's Magazine* enthused, "If ever a philosopher's stone, the elixir of life, the infallible catholicism, the universal solvent, or the perpetual motion were discovered, it is the application of guano in agriculture."

Annual shipments grew to hundreds of thousands of tons per year in the United States and as much in Great Britain. Its availability became a matter of national policy. In 1850 President Millard Fillmore, in his first State of the Union address, noted that "Peruvian guano has become so desirable an article to the agricultural interests in the United States that it is the duty of Government to employ all the means properly in its power for the purpose of causing that article to be imported into the country at a reasonable price." In 1854 twenty thousand Delawareans asked Congress to buy one of the Peruvian guano islands, arguing that "the purchase of one of those islands would be to the people of the United States of more solid worth than that of . . . Cuba and all of the Antilles besides." Diplomatic relations with Peru were, as a result, maintained at a level of extreme cordiality. During the ten years following Fillmore's comment, use in the United States tripled. Farmers, despite steadily rising prices, found it difficult to do without it. Use in Great Britain and France rose almost as fast.

About a third of the sale price of every ton of guano went into the treasury of Peru, a flow of income that by 1859 accounted for three-quarters of the nation's national budget. As one observer wrote, "A stranger means of defraying nearly the whole expenditure of the state was never before heard of." The Peruvian leaders used the enormous income (and the promise of future income) as collateral to leverage foreign loans, fueling a brief gilded age highlighted by a bloated civil service and an inflated military. Wealthy Peruvians built mansions, hired servants, and imported the styles, frills, and trappings of wealthy Europeans. The streets of Lima were decorated with dashing officers in gaudy uniforms and beautiful women in the latest Paris fashions. Government officials involved in overseeing the guano trade, often members of the nation's leading families, became rich from the trade and the bribes that went with it. "As for stealing," wrote a British critic of the trade, "it may be safely said that nearly all public men have steeped themselves to the neck in this crime, and the common people take to it as easily and naturally as birds in a garden take to sweet berries." As long as the money lasted it looked as if Peru, as one enthusiastic observer wrote in 1857, was destined to become "at once the richest and happiest nation on earth."

And then it was over. By the late 1850s the coolies on the Chinchas began to hit rock. The deposits of thousands of years had been stripped away in less than two decades. Traders began looking for the next Chinchas. Any deserted rock with flocks of birds represented a possible fortune. Small guano companies—some of them little more than pirates—scoured the globe, occupying any undefended rock and stripping all the guano they could find.

The need was so great that in 1856 the U.S. Congress passed the Guano Islands Act, which allowed any U.S. citizen to lay claim to any deserted guano island anywhere in the world and make it U.S. territory. The law essentially deputized all Americans to claim land in their nation's name—"an act that has no parallel in history," as historian Jimmy Skaggs put it. Fly-by-night companies began taking territory, including much that should never have been claimed: islands that were inhabited, islands that had no guano, islands that already belonged to other nations, islands they found on old maps, islands they heard about from drunken whaling captains, islands that did not exist. Within a few decades the United States had claimed under the act a total of ninety-four islands, rocks, and keys, many of them sprinkled in that great stretch of the Pacific between Hawaii and Samoa that later became known as American Polynesia. Others were in the Caribbean.

No guano of any quality was gathered from any of these claims. Many of them did, however, serve other purposes.

Midway Island, Baker, Johnston, Howland, and other Guano Islands Act atolls and islets provided airstrips and staging areas in World War II. One island later became a CIA base and was used to beam anticommunist radio into Cuba. One was dubbed "the toxic waste disposal center of the world" because of the nerve gas and other chemical weapons stored there. One, big enough to contain a coconut plantation, was taken over and run for a few decades as a private family kingdom. Eight islands in the Pacific claimed under the act still belong to the United States. The Guano Islands Act is still in effect.

As the Chinchas product declined, entrepreneurs tried to disguise inferior guano. Fast-talking hucksters began selling backwoods farmers sacks of sand and gravel mixed with animal droppings and urine under the label "Pure Peruvian." They mixed poor guano with the leavings from meatpacking plants, or added fish scraps or bone dust, and sold it as "Pendelton's Guano Compound" or "Cotton King Super Phosphate." The port of Baltimore finally had to hire a guano inspector to sniff incoming shipments and assess quality. No amount of ballyhoo or mislabeling, however, could hide the results farmers saw in their next crop. Top-grade guano in bulk, it turned out, was an exceedingly rare thing. In most places around the world, too much rain washed out the nitrogen content and lowered quality. Chinchas-level guano was found on only a very few rocks with the right mix of abundant seabirds and an arid climate. Nothing else came close.

As that fact sank in, as the sources of good fertilizer shrank, the national jockeying for guano became more intense. When Daniel Webster, then U.S. secretary of state, made an ill-fated attempt in the 1850s to lay claim to a few guano-rich rocks off the Peruvian coast, the incident sparked anti-U.S. riots in Lima and the dispatching of Peruvian troops to the barren islands. The United States backed down.

A real war started in 1863 when a Spanish naval squadron occupied the Chinchas in a move both ill advised and ill timed.

The end of Chinchas guano was already in sight, and the Spanish occupation—made by an arrogant and hot-tempered admiral—kicked off declarations of war by both Peru and its neighbor Chile. The two former colonies combined their navies under the command of a talented former Confederate admiral from Virginia and succeeded in capturing a Spanish ship. The humiliated Spanish admiral committed suicide and his nation withdrew. The "Guano War" marked South America's true independence from Spain and set off an exuberant two-nation celebration.

Toward the end of the guano trade "blackbirders"—pirates who kidnapped Polynesian and Melanesian tribesmen—raided the Easter Islands, shipping a thousand captured tribespeople to work digging guano. They included about a third of the islands' total population, including its chief, crown prince, and most of the priests. It is estimated that nine hundred of the captured Easter Islanders died in captivity. When the raid became public, an international outcry led to the attempted return of the one hundred or so survivors. Eighty-five of them died of disease on their way home. The remainder passed on the disease when they finally arrived home. By 1877 only 111 natives remained in the Easter Islands.

By then the Chinchas were bare. "When I first saw them twenty years ago they were bold . . . tall and erect, standing out of the sea like living things, reflecting the light of heaven or forming soft and tender shadows of the tropical sun on a blue sea," wrote a nostalgia-addled guano man at the time. "Now these same islands looked like creatures whose heads had been cut off, or like vast sarcophagi, like anything in short that reminds one of death and the grave." That was in the 1860s. Today the Chinchas again belong to the birds. The Peruvian government restricts access, protects the wildlife, and still allows some limited gathering of what is still the world's best natural fertilizer.

The end of guano also meant an end to Peru's easy income.

By the 1870s the nation was for all practical purposes bankrupt. About eleven million tons of guano had been gathered, shipped, and spread during the brief guano age. Now that it was over government officials in Peru faced disaster and farmers in Europe and America, increasingly hooked on applications of high-quality fertilizer, were facing a return to the bad old days of declining yields.

It was then that businessmen, farmers, and politicians remembered that there was another natural source of fertilizer just south of the Chinchas. It was called *salitre* by the locals in the Tarapacá desert; the traders in Britain called it nitrate. There were hints in the reports of visitors like Darwin that the nitrate might exist in great quantities, covering the ground for miles. There were reports from a few farmers that it could do wonders for crops. At the same time the leaders of Peru realized that this new source of wealth also belonged almost entirely, by the grace of God, to their nation.

For the moment.

CHAPTER 4

SOUTH AMERICA'S GREAT Atacama Desert is a place unlike any other. Its climate is different, with close to zero rainfall but occasional thick fogs. Its plants and animals are different—what there are of them, which is to say almost none—capable of living with almost no water. Even its rocks are different. The floor of the Atacama is crusted and shot through with a riot of strange chemicals: nitrates, chromates, and dichromates; perchlorates, iodates, sulfates, and borates; chlorides of potassium, magnesium, and calcium; minerals "so extraordinary," a researcher wrote, "were it not for their existence, geologists could easily conclude that such deposits could not form in nature."

How they did form is still a question. It is generally agreed that the area's dryness probably has something to do with it, that anywhere else in the world these minerals would not have had the chance to aggregate because they would have dissolved in the rain or been eaten by microorganisms or absorbed by vegetation. Because there is no rain in the Atacama, there is no life. There is some thought that the fog might play a part, or perhaps the nearness of the sea, or salts washed down from the Andes. Whatever the reason, the Atacama is the only place on earth that significant deposits of these rare minerals slowly crystallized, century by century, to form the nitrate-rich mineral layer the natives call *caliche*.

The richest layers of *caliche* are found in a strip along the western edge of the desert, just inland from the sea, over the hills Darwin crossed. This ribbon of desolation, five to ten miles wide and a few hundred miles long, contained almost all the naturally occurring sodium nitrate on earth, stored in layers of *caliche* as thick as a man was tall.

In the 1870s, as the guano disappeared, the world began to appreciate its value, both as a fertilizer and as a component of fireworks. European and U.S. merchants bought it in small quantities for both purposes, and sometimes got them mixed up. One early attempt to import nitrate into England for fertilizer failed, the story goes, when Liverpool officials refused to accept it because the danger of explosion was too great. The captain ended up dumping it in the water. It took time for European and American farmers to appreciate its benefits, especially when all their attention was on the marvelous fertilizer from the Chinchas. In the early part of the nineteenth century, bags of the desert nitrate were sometimes used primarily as ballast for returning European ships, a way to put weight in their bellies and possibly make a little profit on the side.

The dynamic changed for two reasons: The guano boom ended, and a chemist figured out how to turn the South American nitrate—suitable only for making inferior explosives—into true saltpeter, China Snow, the world's most important component of top-quality gunpowder. Only a single atom differed between South American nitrate and true nitrate, and when chemists figured out how to switch them, cheap Atacama nitrate suddenly grew in value. Making South American nitrate into true saltpeter was only the beginning. The newly made true saltpeter could then be chemically altered again to make tons of nitric acid, the basic ingredient used in making a new generation of "high explosives" like nitroglycerine (developed by Alfred Nobel in the 1860s) and dynamite (patented by Nobel in 1867). As the 1800s went on, increasing amounts of the Atacama products were shipped to Europe to make not better crops but better explosives, including those used in high-explosive shells for the military. The fact that Atacama nitrates could be used for both food and war made it, in the latter half of the nineteenth century, one of the world's most valuable natural resources.

For years Britain had the upper hand when it came to gunpowder, thanks to its control of the true saltpeter deposits in India. During the Crimean War in the 1850s, however, the increased need for explosives sparked Britain's interest in the Atacama. The South American nitrate trade began to take off, with Britain soon joined by Germany and France, all three nations investing in nitrate shipping, the purchase of nitrate-rich Atacama land, and the construction of refineries in the desert. The United States began boosting its nitrate trade in the 1860s to make gunpowder during the Civil War. Peru, which controlled the Atacama, began to make big money from the trade.

Everyone started staking claims in the desert. The growing number of nitrate refineries in the Atacama drew thousands of Chilean peasants, tough, independent-minded, hardworking men and women who trekked north to Peru to trade their lives as agricultural workers—peasants—for the promise of steady, higher wages. Nitrate ports mushroomed along the coast, populated by prospectors, salesmen, water brokers, suppliers, prostitutes, British businessmen, German engineers, French shippers, and American adventurers. The richest diggings were in the northern Atacama, the district of Tarapacá, where under Peruvian law almost anyone could claim about ten acres of desert land and set up a nitrate refinery as long as they paid the government fees and taxes on what they produced. Families started staking as many claims as they could.

Making the *salitre*—the refined nitrate—was difficult, dangerous work. The salts in their raw natural form, *caliche*, were often buried below a layer of dirt and rock, so prospectors called *barreteros* were hired to find the *caliche*, digging trenches to locate the salt layer, testing it for purity (the best deposits were around 50 percent pure nitrate), and deciding if it was worth processing. Quality testing in the early days consisted of grinding up a bit of *caliche* and throwing it on a fire to see how well it burned. The *caliche* was as hard as rock. It had to be blasted with explosives, and the *barretero* was often in charge of this as well. Then the *calicheros* came in, the workers who shoveled away the dirt, dug out the *caliche*, pickaxed it into handleable pieces, loaded it on mules, and hauled it to the refinery.

By 1863 there were nine big, new steam-powered refineries, locally called *oficinas*, operating in the Tarapacá, employing hundreds of workers to pulverize the *caliche* between iron rollers, boil it in big iron vats, purify the nitrate, then bag, load, and transport it. Production rose, exports doubled, then tripled, then went through the roof in the 1870s as the world ran out of guano and began to realize that there was not only fertilizer but also gunpowder in the desert. The *caliche* was so rich, the Atacama so big, that the supply seemed endless.

As the long bender of the Guano Age ended, the Peruvian government sobered up and began to realize that its future was in nitrate. By then, however, most of the refineries, most of the production, most of the transportation, most of the brokering, and almost all of the shipping, were in foreign hands, Chilean, British, German, and French. By the end of the 1870s more than twenty-five thousand Chilean laborers had entered Peru to work the nitrate. By the time Peru realized what it had, the whole area had been essentially colonized.

But that made little difference to the leaders in Lima. What counted was not ownership but the money that could be made to flow to the Peruvian government. In the wake of the guano collapse, Peru needed money. It had floated enormous foreign loans on the promise of continued guano production; by the late 1860s Peru was one of the world's largest debtor nations. It needed whatever it could make out of nitrates as quickly as possible. So, just as they had done with guano, the Peruvians allowed foreigners to do the dirty work while the government raked in money from taxes, fees, and licenses.

Iquique, which in 1869 was still "little more than a shabby collection of unpainted frame buildings scattered along unpaved streets," as one visitor described it, had by 1871 been transformed into a frontier boom town with almost twelve thousand inhabitants. It was the Tarapacá's biggest city, the seat of local Peruvian government and home to swarms of new inhabitants who lined up at notary offices to lay land claims, fight over building sites, and try to corner markets on imported water and exported nitrate. Iquique became famous for its rough life, its loose women, and its alcohol consumption. "The West Coast [of South America] was a veritable paradise for drunkards," one historian wrote. Sailors set free after weeks at sea found little to do but buy a few *treepas*—long, sausage-shaped skins filled with wine, three for a dollar—and go looking for trouble. One appalled English observer wrote of the typically "ignorant, drunken, and grossly depraved" English sailor in Peru: "When on shore, if he is not drunk, he's kicking up a row. His language is foul. His manners brutal ... and his appearance that of a wild beast." The sailors seemed to enjoy it. A number of them decided to trade life aboard a British merchant ship for life in Iquique and deserted, finding plenty of chances for work in the boom town, or joining the Peruvian army or navy, or simply drinking and combing the beach.

Soon the best deposits were stripped. The early *oficinas*—small mills run by families for a few years until the richest nearby *caliche* was gone, then dismantled and reassembled at the next good diggings—gave way to larger, more efficient, more permanent factories. Even worked-over areas of the desert still contained enough nitrate to make money. It only required better methods to refine it. These incongruous, hulking mills in the blankness of the desert—sprawling, steaming complexes built with imported Oregon timber, fitted with European machinery, and served by railroad lines—turned into small towns with hundreds of workers, stores, maybe a theater, a tavern, a brothel, a hospital, and their own brass band. "[A]n uncountable number of vast iron tanks containing colored liquids," as a visitor described a nitrate refinery in 1877, "a tall chimney, a chemical laboratory, an iodine extracting house, a steam pump, innumerable connecting pipes, stretching and twisting about the vast premises as if they were the bowels of some scientifically formed stomach of vast proportions . . . a blacksmith's forge, an iron foundry, a lathe shop, complicated scaffolding, tramways, men making boilers, men attending on wagons, bending iron plates, stoking fires, breaking up *caliche*, wheeling out refuse, putting nitrate into sacks . . . all beneath the fierce heat of a sun . . . as painful as looking into a blast furnace." The desert was being torn up and consumed, the vast, empty Atacama transformed in a few years into what one visitor called "a vast number of ant-hills."

But even the new flood of nitrate money was not enough to save Peru. Its government, lurching from crisis to crisis, finally decided in the mid-1870s that it should take over the nitrate industry, nationalize it, and rake in even bigger profits. The refinery owners, mostly European and Chilean, were dead set against it and considered armed insurrection rather than giving up their business. The large number of Chileans working Peru's mills added another dimension to the issue. In 1874 an attempted revolution was funded and armed in part by Chileans who, it was said, wanted to use the unrest to take over some of Peru's nitrate holdings. It failed. But the nitrate producers began thinking hard about a way out from under the Peruvian government. Increasingly, they looked for help to the south, to Chile.

The government of Peru, an observer of the day wrote, "is and has been from the commencement of its Republican history, as unstable as water." Chile was a different story. The two nations, close geographically and tied by a common language, differed in almost every other way. Peru, home of almost all the easy wealth, the gold and silver of the Incas, had as a result received most of the attention from Spain. The Spanish conquistadores had turned it into a vassal nation, raped it, and looted its riches. It became even after independence a nation of a few very rich families ruling many extremely poor citizens. Chile, by comparison, had many fewer Indians, far less natural wealth, and a heavier influence of settlement by German and British immigrants. The Peruvian tradition was one of aristocratic control, slave labor, and resource exploitation. The Chilean tradition was relatively progressive, nationalistic, enterprising, and aggressive. The Chilean government was stable. Chileans were compared by some historians to the American pioneers.

The two nations had been allies, briefly, during the Guano War with Spain, but traditionally acted more like competing brothers. The competition began to turn into a brawl when they both saw something they wanted: the desert nitrate. As complaints from Chileans working in Peru increased, the two nations began gauging each others' military forces. "If Chile buys one warship," went an oft-quoted dictum of the era, "Peru must buy two." There was a third player in the region too, Bolivia, which controlled a strip of nitrate desert between Chile and Peru, including a nitrate port. Chile had a longstanding dispute with Bolivia over where, exactly, one nation ended and the other began. Twenty years earlier the dispute had been about little more than a small point of national pride. Now it was about big money. Peru, alarmed by Chile's growing naval strength, urged Bolivia to reject any Chilean claims. Peru and Bolivia signed a secret mutual defense pact.

The nitrate war started in 1879.

THE BOLIVIANS MADE good targets. Their uniforms were the colors of the mountains and jungles, bright green and yellow and red, easy to spot against the dull sands of the Atacama. They had marched down from their homes high in the Andes under the command of the nation's newest dictator, General Don Hilarión Daza, an "adventurer of the lowest and worst type," a man whose military abilities were no match for his greed. To stay in power Daza needed to keep his army happy, so he rifled the Bolivian treasury to pay his officers and men. He needed jobs to keep his people from revolting, so he arranged "flexible" (often graft-driven) deals with foreign firms that would employ Bolivians in the mining of their abundant silver. The Daza regime, installed in fortresses high in the mountains, was renowned less for its good works than it was for its corruption and much-whispered-about palace orgies.

When he heard that money was being made in his nation's distant holdings in the Atacama, from the mining of a valuable salt, Daza ordered a small increase in the taxes being paid by the industry. (Most of the work in Bolivia, as in Peru, was being done by Chileans.) His advisers told him that this was not possible: One of Daza's predecessors had brokered a deal with Chile in which a border dispute had been settled in exchange for tax breaks for Chilean settlers, including a promise not to raise taxes on the nitrate works. Daza, however, would not be bound by the negotiations of the past. Daza decided to levy his tax anyway.

It was not much. Just enough to kill a few thousand soldiers.

Enterprising Chileans in the Bolivian Atacama had already turned a fishing village into a major nitrate shipping port. In the late 1870s the population of this fast-growing Bolivian city, called Antofagasta, was 90 percent Chilean. When Daza levied his tax, the manager of the region's biggest nitrate company—run by Chilean and English businessmen—refused to pay. Daza responded to this affront by taking over the company in Bolivia's name. Chileans in the area called for help from the government of Chile.

On February 14, 1879, without warning, two hundred Chilean troops arrived by ship in Antofagasta harbor and marched up the dock through lines of cheering residents. The Chileans did not want bloodshed; they quickly took control of the town and gave the town's Bolivian prefect and all other Bolivians free passage back to their homes in the mountains. When news of this insult reached Daza, he declared war on Chile. He was quickly joined by his secret ally, Peru.

Daza personally led his Bolivian soldiers. It was a mistake from the start. Daza was as poor a general as he was a politician. His troops were soundly beaten by the Chileans whenever they met. Within weeks Daza fled back to the Andes with what troops he had left, leaving a relatively small Chilean military force in control of almost all of what had been the Bolivian Atacama. It was a humiliating defeat. Bolivia lost its nitrate holdings and its arm to the sea (it remains to this day a landlocked nation) and Daza was ousted. He fled with a chunk of the national treasury first to Europe, then to Peru. When he tried to return to Bolivia a few years later, he was shot to death at the border.

He had, however, succeeded in one thing: He had started what was called the War of the Pacific, or the nitrate war. At stake was the entire Atacama, including the Tarapacá's rich nitrate fields and all the fortunes being made there, a natural resource important enough to change the fate of nations. It was a prize worth winning.

THIS WAR FOR the desert was fought largely on the sea. In a land as dry and endless as the Atacama, only the sea offered a way to move troops and arms quickly and safely. Control of the sea meant control of the war.

On paper, Chile looked to have the advantage—its ships were newer and faster—but Peru had two superior weapons.

One was an aging, slow, but almost impregnable ironclad, a floating fort sheathed in four inches of iron plating called the *Huáscar*. The other was that ship's captain, Miguel María Grau Seminario, the former director of the Peruvian naval academy. Grau was a talented naval strategist. He knew that the *Huáscar* was not the fastest or most up-to-date ironclad afloat. But he also knew that it was a heavyweight in any fight, with armor thick enough to repel enemy cannonballs, two huge cannons housed in a squat revolving tower, and an iron beak—a ram— that could slice a hole in the enemy and sink them at close quarters. One on one, Grau felt, he could sink or capture any ship in the Chilean navy.

Both sides focused on the port of Iquique. Darwin's gloomy fishing village now sported new docks and an imposing stone Peruvian blockhouse on the waterfront. As soon as the war started, the Chileans blockaded Iquique, denying Peru money from nitrate shipments and waiting for a chance to take the town. Grau, the Peruvian naval commander, determined to break the blockade. A cat-and-mouse game ensued. The Chileans were rightly nervous about the *Huáscar* and its captain, so they sent some of their blockade ships north from Iquique toward Lima to find and sink Grau. But he was already steaming to Iquique in the *Huáscar*, accompanied by Peru's only other ironclad in good enough shape to send to sea, the *Independencia*. Somehow he got to Iquique without being seen. There were only two Chilean ships maintaining the blockade.

What the Chileans lacked in power, they made up for in courage. One of the Chilean ships, the corvette *Esmeralda*, was commanded by a handsome thirty-one-year-old officer named Arturo Prat. When the *Huáscar* was sighted at dawn, Prat, according to local lore, went below, put on his parade uniform, and roused his men with an impassioned speech. "Lads," he said, "the battle will be unfair, but, cheer and courage. Our flag has never been hauled down before the enemy and I hope this will not be the occasion to do it. For my part, I assure you that as long as I live, this flag will blow in its place, and if I die, my officers will know how to fulfill their duties."

Grau went after Prat's ship. The other Chilean tried to escape with the Peruvian *Independencia* giving chase. The *Huáscar* trapped Prat against the coastline and started hammering the *Esmeralda* with its big guns. By now crowds of cheering Peruvians from Iquique were lining the waterfront, welcoming Grau, bringing out what cannon they had, and taking potshots at Prat's ship. Running out of room, Prat repositioned the *Esmeralda* so that the *Huáscar* would have to fire at an odd angle to avoid hitting fellow Peruvians on shore. For the next hour and a half the two ships traded blows. The Chilean fire was more accurate but their cannonballs bounced harmlessly off the *Huáscar*'s iron plating. A cannonball from the *Huáscar* killed the Chilean surgeon and beheaded his assistant. Then the Peruvians on shore began to hit the *Esmeralda* with cannon fire. Prat tried to reposition but his engines started failing. Then a boiler burst and the *Esmeralda* was dead in the water. For the next hour and more the *Huáscar* continued smashing the crippled Chilean ship, disabling its cannons and softening it up for a final approach with the ram.

Perhaps Prat should have surrendered, but he had promised his men that he would never lower the Chilean flag. Grau ordered the *Huáscar* to back up a bit, then steamed forward and smashed into the *Esmeralda*. The *Huáscar*'s ram tore open the Chilean ship below the waterline. As his ship shuddered from the blow, the two ships deck to deck, Prat raised his sword and ordered his men to board the *Huáscar* and take it hand to hand. Perhaps his crew did not hear through the noise, perhaps the sight of the *Huáscar*'s Gatling gun sweeping the deck dissuaded them. Whatever the reason, only Prat, still brandishing his sword, and a single sergeant armed with a pistol and a boarding hatchet made it onto the *Huáscar*'s deck. Within seconds the sergeant fell, mortally wounded. Prat ran across the deck alone toward the bridge of Grau's ship. The Peruvian officers were, for a moment, too stunned to speak. Then Grau issued an order to take the enemy captain alive. But it was too late. There was a shot and Prat fell to one knee. There were more shots, and he died on the *Huáscar*'s deck.

Grau ordered his men to carry Prat's body below. Then he ordered the *Huáscar* back and rammed the *Esmeralda* again. This time Prat's second in command tried to board the Peruvian ironclad, and ten Chileans made it onto the *Huáscar* before the Gatling gun cut them down. The *Esmeralda* was sinking, its deck almost awash, when the ironclad rammed a third time. It finally sank, still flying the Chilean flag. One hundred forty eight of the *Esmeralda*'s 198 crew members were dead. The *Huáscar* lost one sailor.

An admiring American historian later called Prat's final minutes the greatest act of bravado in a naval battle since the days of John Paul Jones. Grau was deeply impressed as well and sent Prat's young widow all of his personal belongings, including his sword, with a note:

Dear Madam:

I have a sacred duty that authorizes me to write you, despite knowing that this letter will deepen your profound pain, by reminding you of recent battles.

During the naval combat that took place in the waters of Iquique, between the Chilean and Peruvian ships, on the 21st day of the last month, your worthy and valiant husband Captain Mr. Arturo Prat, Commander of the Esmeralda, *was, as you cannot ignore any longer, victim of his reckless valor in defense and glory of his country's flag.*

While sincerely deploring this unfortunate event and sharing your sorrow, I comply with the sad duty of sending you some of his belongings, invaluable for you, which I list at the end of this letter. Undoubtedly, they will serve of small consolation in the middle of your misfortune, and I have hurried in remitting them to you.

Reiterating my feelings of condolence, I take the opportunity of offering you my services, considerations and respects and I render myself at your disposal.

Both sides claimed victory at Iquique. The Peruvians sank the *Esmeralda*. The Chileans gained a national hero in Arturo Prat (the town square and leading hotel in Iquique are now named after him). Chile also benefited from a lucky accident: While chasing the second blockade ship, Peru's *Independencia*, that nation's only other seagoing ironclad, ran into a reef and sank.

For several months after the battle of Iquique, Grau and his impregnable *Huáscar* ruled the seas off Chile and Peru, the ironclad destroying or capturing every Chilean ship it could catch, bombarding ports, preventing Chile from moving men by sea, and sinking Chile's hopes of a quick victory. Grau moved slowly but stealthily, struck by surprise, and always seemed one step ahead of the Chileans. He was promoted to grand admiral of Peru's navy. His countrypeople dubbed him *el Caballero de los Mares*: the Knight of the Seas.

By the fall of 1879 Chile realized that its only hope of victory had to start with the sinking or capture of the *Huáscar*. Six of Chile's best warships were devoted to the task. They finally met Grau on October 8. What shaped up to be a major sea battle ended quickly when a lucky Chilean shot hit the *Huáscar*'s bridge, killing Grau and several of his officers. Two hours later the *Huáscar* was captured.

As the news spread, Chilean crowds danced in the streets, chanted insults, called the Peruvians cowards and guinea pigs (in Peru, guinea pigs are roasted and eaten), and shouted, "On to Lima!" The loss of one ship changed the entire nature of the war. With the *Huáscar* captured and Grau dead, the Chilean navy controlled the Pacific from one end of South America to the other. They quickly captured Iquique, started shipping soldiers north, and prepared to take the war to Peru's heartland.

Fifteen months after losing the *Huáscar*, the Peruvians had been beaten on every front. Now they were trying desperately to stop the Chilean forces from capturing their capital city. An American on the scene wrote, "I followed the thoroughly disorganized [Peruvian] army into Lima. Here we found the city in the hands of a mob, which plundered and burned to their hearts' content. Bullets flew through the streets as thick as hail; the yells of wounded men, the shrieks of the frightened women, the red glare of a hundred fires, all went to form about as perfect a picture of hell as one could well imagine."

In the end, he wrote, it came as something of a relief when the Chileans marched in and restored order.

CHAPTER 5

Chile's victory was total. In exchange for Lima and a bit of its pride, Peru ceded to the victor all of its nitrate-rich desert, including the Tarapacá, as spoils of war. This meant that after 1881, Chile had sole control over the world's only significant deposits of the world's most valuable natural resource. It would be as if, today, a single nation controlled all the oil wells in the world.

Chile was about to become rich.

Europe and the United States were by now dependent on nitrates to grow their food and arm their military. By 1900 the United States was using almost half its rapidly increasing tonnage of Chilean nitrates to make high explosives, giving it the power to push through railroad lines, deepen rivers, dig mines and tunnels, level roads, carve Mount Rushmore, and blast through the Panama Canal. It was made possible by Chilean nitrates, tons upon tons of them, the shipments from the Atacama increasing year by year. The more demand grew, the more the Chileans produced.

Explosives were especially important in a United States busily securing its Manifest Destiny, but just as important—and more important to most European importers—was the use of Chilean nitrate to grow crops. The world population rose rapidly between 1880 and 1900. As workers flocked to fast-growing cities, fewer farmers had to feed more and more factory workers. Diets were changing too, in the most industrialized nations, shifting away from vegetable-and-grain subsistence toward richer, more highly processed foods, more meats, sugars, and oils. All of this added up to a dramatically increased need for higher agricultural yields, which again were made possible by enriching the soil with Chilean nitrates. By 1900, Chile was producing two-thirds of all the fertilizer used on earth.

Among the world's biggest buyers were Great Britain and Germany, two nations where land was limited but military aspirations were not. Britain at least had a world-spanning empire to help grow its food, a commonwealth that stretched from the grain fields of Australia and Canada to the true saltpeter deposits in India. Germany, on the other hand, was formed too late (created in its modern form in the 1870s) to get in on the age of colonial expansion. Without many colonies, Germany had to grow most of its own food on its own, generally poor, soil. As a result, it became by far the biggest user of Chilean nitrate by the turn of the century, importing more than 350,000 tons a year in 1900 and more than 900,000 tons in 1912—twice the amount imported into the United States and three times that shipped to France.

Nitrates were Germany's life blood, and their distance from Germany—halfway around the world—became a matter of strategic importance. German entrepreneurs created fleets of majestic windjammers for the nitrate trade, the biggest sailing ships ever built, with clouds of sail that drove them so fast that they were preferred even in the age of steam. The master builder of nitrate windjammers was the Laeisz company in Hamburg, whose Flying P Line ruled the trade for decades. Laeisz hired the world's best captains and crews, maintained meticulous discipline, developed good relations with port officials, and made the best possible deals with companies who ran the "lighters" (smaller boats that carried sacks of nitrate out to the massive clippers). It took most ships about three months to sail from Iquique to Europe; a Laeisz clipper once did it in sixty-five days. Many nitrate ships spent two or three months in port in Chile; Laeisz ships were in and out in less than two weeks. They turned nitrate shipping into a science.

Chile kept up with the rising demand by building more and better refineries in the desert. In the twenty years after Chile's victory in the War of the Pacific, the number of mills more than doubled, the number of workers tripled, technologies improved, shipments mushroomed, and profits skyrocketed. Soon nitrates were bringing in more than half of Chile's total income (nitrates would remain the single most important factor in the national economy until the 1930s). Chile did not waste its income, as Peru had wasted its guano riches, but invested instead in telephone and electrical systems, transportation networks and schools, a bigger government, a more professional army, and, of course, a top-of-the-line new ironclad for the Chilean navy. Nitrates paid Chile's entry fee into the modern world.

For two decades, from the end of the nitrate war until the night Crookes gave his speech, no one spoke of a shortage of nitrates. Better refineries kept wringing more out of the diggings, the desert kept producing, and shipments increased year by year. The world demand for nitrates kept rising. Money kept flowing. People acted as if it would never end.

JOHN THOMAS NORTH, age twenty-nine, debarked in Iquique in 1871. He was one of thousands of British citizens who came to seek their fortunes in South America during the nitrate era. He had some training as a mechanic and a keen eye for business, and he started his new life in South America by getting into a surefire business: the shipping of water to Iquique from springs farther north. He familiarized himself with the nitrate trade as well, watching as desert properties were bought and sold, and speculating a bit himself if he saw an attractive deal.

Then came the war and the Chileans. The takeover of

Iquique and the Tarapacá threw all previous ownership—all the deals that Peru had made—into question. In the confusion and uncertainty many owners were willing to sell cheap. North began buying. He worked closely with the new Chilean inspector of nitrate mills (another British-born émigré) and was soon making serious money, buying nitrate properties at distress-sale prices and turning them around, sometimes in the same day, for two, three, or four times as much. He advertised his success in London, where he used his growing personal fortune and portfolio of nitrate holdings to float loans to start companies like the Liverpool Nitrate Co. Ltd., which gathered British investors for more purchases. In the mid-1880s, nitrate speculation went wild in England, thanks in great part to North's enthusiasm. During his increasingly opulent visits home, he described to the rich of his nation the enormous profits to be made by anyone smart enough to buy and sell chunks of the desert and shares in the refineries that made nitrate. It was an absolutely necessary commodity. Fortunes were there, in Chile, for the taking. Enormous investments were made in North's companies. North built some refineries himself, and then sold stock in them. His investors began making 10, 15, 20 percent dividends in a single year. "The company promoter has only to whisper the magic word 'nitrates' and the market rises at him," marveled London's *Financial News*. Speculators started calling Chilean nitrate white gold.

The man who had gotten off the boat in Iquique in 1871 was by 1882 "Colonel" North, England's leading nitrate speculator and one of its richest men. His interests grew to include railroads and more mills; it looked for a while as if he were going to buy up the entire nitrate region. When in Chile he toured his holdings in a palatial private railroad car. When in England he lived in an enormous mansion he built outside London. He was introduced to the Prince of Wales. His star reached its zenith in January 1889, when North and his wife hosted a fancy-dress ball for eight hundred at the Hotel Metropole in Whitehall Palace. Present were Sir Randolph Churchill, Baron Alfred de Rothschild, thirteen peers, fifteen knights, and the sheriff of London. North dressed as Henry VIII, his wife as the Duchesse de Maine.

Then the bubble burst. Rampant nitrate speculation led by North pushed share prices well beyond the value of the land and mills. To keep returning high profits the refineries overproduced, flooded the market with nitrates, and held down prices. By the end of 1890 shares in nitrate companies were being sold for a quarter the price they had commanded a year or two earlier. One day North's broker entered the London Exchange to find men piling copies of his latest company's prospectus on the floor and setting them on fire. North kept telling nervous investors that nitrate sales naturally fluctuated, that the core value of his enterprise was buttressed by the value of his railroads alone, and that the long-term future was bright. But even North could not single-handedly keep the nitrate market afloat. As precipitously as they had gone up, the value of nitrate investments went down. North, his fortune depleted but not gone, retreated into private life.

THROUGH IT ALL, Chile maintained its profitable and well-ordered business in nitrates. Producers knew what was expected, paid their fees and taxes, and reaped their profits. A reasonable portion of what came in went to the Chilean treasury and was used for the betterment of the nation. The industry retained enough to keep its investors happy and to pay for continuing technological advances for the refineries, which were by 1900 many times more efficient than they had been a few decades earlier, able to profitably work nitrate fields that earlier producers would have scorned. Chile, like Peru, depended on foreign capital and foreign expertise (mostly European) to power those improvements. Unlike Peru, however, it kept its own people employed in the industry and offered a stable atmosphere for development.

A number of companies were Chilean owned, but by no means the majority. Nitrate ports like Iquique became surprisingly international. There were English mill owners (by 1890 British companies owned 70 percent of the nitrate mills in Chile); German engineers; Chilean businessmen, government officials, and laborers; Spanish, Italian, Bohemian, and French shippers, shop owners, nitrate brokers, and hoteliers; Indians; a few remaining Peruvians; and a community of former Chinese coolies. Iquique, an isolated, waterless fishing village a few decades earlier, was now incongruously cosmopolitan, with a grand central square, elegant hotels, and a new opera house. It was a place where bon vivants could entertain themselves at a German tavern or a British club, eat at a Chinese restaurant, attend an opera given by a visiting French troupe, and top it off at an Italian café. There were cricket matches and soccer games, shooting parties, elegant balls, and evenings at the theater.

That was life along the coast. In the desert, in the nitrate factory belt, life was far different. The growing number of mills drew thousands of poor laborers, many of them former farmworkers from southern Chile, lured north by the promise of easy money. In a typical nitrate mill an entire family would live in a twelve-by-twelve-foot room in a complex of long, low, barrackslike buildings near the works. The men worked six days a week in the blazing sun blasting holes in the desert, digging and hauling *caliche*, operating crushers, tending boilers, stoking fires, scraping vats, sacking nitrates, and hoisting three-hundredpound bags onto railroad cars. For this they received their three pesos a day—but not in the form of money. Most nitrate companies paid their workers in *fichas*, a kind of company scrip redeemable at only company stores. It was as if each nitrate company minted its own money. The *ficha* system ensured a steady supply of customers for stores that often charged high prices, but its most important effect was to keep a company's workers in place. Trained laborers were valuable. Jobs in the desert were plentiful. If workers could not spend their wages anywhere else, it was unlikely they would move anywhere else.

To the workers, the *ficha* system was a cage, a chain, and a collar. Not only were they charged high prices for every bite of food and drink of liquor, not only were their wages useless in any other town, but the *fichas* also made saving money useless. Saved *fichas* could not be spent anywhere else. In many ways, getting paid in *fichas* was like not getting paid at all.

So the workers complained to each other, drank together at the company *pulperias*, and slowly developed a sense of shared outrage, a sense of being used—and a sense of unity. The men began to listen to Chilean labor activists who talked about communism and anarchism and workers' rights. The women cooked, tended children, shared chores, shopped together at the company market, and stared out at the silent, endless desert. Then they, too, began to talk.

Their unhappiness was leavened with feelings of pride. The *calicheros* and *salitreros* of the desert, the nitrate workers, saw themselves as unique, strong, among the hardest-working people on earth, and important to their nation. Chile's great poet Pablo Neruda spent time in the nitrate fields and ports and understood the breed. He wrote of port workers shoveling *salitre* into the holds of ships as "heroes of a sunrise," creatures of the desert with bodies covered in sweat, chests penetrated by acids, their hearts swollen "like crushed eagles, until the man drops."

The mill workers began organizing, some forming radical theatre groups that traveled the desert singing songs and satirizing mill life, others setting up *mancomunales*, mutual-aid associations, early steps toward forming more powerful labor groups. It was the beginning of a long tradition of left-wing political action in Chile.

• • •

THE WORKERS CAME, singing, down the long road out of the desert of the Tarapacá and into Iquique in the summer of 1907. There was a mass of them, more than anyone had ever seen in one place before, agitating for an end to the *fichas*, for better wages, for better lives. The worried townsmen allowed the first workers to camp in the yard of the Santa Maria school. But they kept coming over the next several days, thousands upon thousands, pouring out of the desert.

Labor demonstrations had started in earnest in the nitrate fields around 1903, tied to the growth of anarcho-syndicalist resistance societies across all of Chile for all sorts of laborers. Small strikes were used in the beginning, perhaps a daylong work stoppage at a mill when the men and their families would gather to hear speeches in the plaza. Almost always the mill owners would refuse to bargain, and the government would step in and get the laborers back to work.

The 1907 march started the same way, with workers at a nitrate mill near Iquique laying down their tools. This time, instead of staying at the mill, they decided to take their protest into town. At first it was peaceful. The people in Iquique were interested in the speeches in the schoolyard. A local girl caught the moment in a letter to her grandmother:

> *There were many: men, women and children, grand-mothers and grandfathers. They also had their dogs that ran in between their legs as if knowing that they participated in an important event. The women came with baskets, pans and spoons, the babies against their breast, and the men with their younger children on their shoulders. It was very hot those days. The* camanchaca *[fog] did not bring its usual relief. The heat hovered over the city like a heavy mantle. The days passed and in spite of the quantity of people, there was*

an air of hope. According to Juan, the people from the pampa said they would wait until their petitions were accepted. They wanted to change many things, Grandma, such as for example, eliminate the fichas *and have schools in the afternoon and better medical attention.*

It is still difficult to know exactly what happened next, because eyewitness accounts differed dramatically. It is known that around 1900 the total population of Iquique was about sixteen thousand. After a few days, the number of protesters around the Santa Maria schoolyard grew to somewhere between six thousand and twenty thousand workers and family members, depending on whether you use the figures given by newspapermen, government officials, or labor activists. It appears they were entirely peaceful. A strike committee was set up, and people spent their time talking, educating one another, and planning next steps. There was something of a holiday atmosphere; instead of backbreaking labor, the strikers got to rest, eat, and listen. The strikers were impressed by the power of their own numbers. For a few days there was a sense that the sheer presence of so many workers, plus a firm refusal to leave Iquique until they were heard, would result, finally, in the end to the *fichas*, or better wages, or some kind of movement by the owners.

Iquique town officials notified the government in Santiago, requested help to maintain order, and tried to mediate agreements between the management and the workers. But talks fell through after the mill owners flatly refused all demands. Chile's president tried to step in, offering to pay half the raise demanded by the workers if the owners contributed the other half, but that, too, was rejected. Now the townspeople began to worry. How were the workers to be fed? They were drinking all the town's water. There were no big problems yet, but if any were to arise, Iquique's few police would be overwhelmed. The holiday mood began to evaporate.

Two Chilean ships arrived carrying a military force under the command of a tough-minded general, Roberto Silva Renard. The Chilean government was alarmed about the size of the desert protest, the refusal of the workers to return to their mills, and especially about the example this provocation would set for workers in the rest of the country. They did not want the Iquique gathering to spark a nationwide workers' revolution.

General Silva was ordered to end the strike. Silva had machine guns from the ships set up at the schoolyard. Soldiers surrounded the workers. Martial law was declared.

On the afternoon of December 21, 1907, the longest day of the year in Iquique, Silva gave the strikers a deadline to disperse. Instead of leaving, the strike committee called a meeting on a school balcony under the flag of Chile. The workers in the schoolyard stayed with them. The general's deadline passed. Then, shortly afterward, the soldiers surrounding the schoolyard were ordered to ready their arms. Some of the soldiers, especially some who had grown up in Iquique, refused to raise their guns. Silva turned to men from the ships, men who came from other parts of Chile, the sailors manning the machine guns. They were ordered to fire.

The strike leaders under the flag were shot first. "As though struck by lightning they fell," wrote a witness. The crowd panicked. It was reported that some of the strikers had weapons and might have tried to use them. The machine gunners strafed the schoolyard, firing for what seemed like a few minutes, no one was quite sure. There were cries and screams, there were many women and children. The crowd broke and ran; the schoolyard emptied. Only bodies were left behind. Iquique officials claimed that 126 strikers died; the British minister in Santiago estimated somewhere around 500. Historians now believe that somewhere between 1,000 and 3,000 workers, spouses, and children were killed. There were no reports of any military casualties. The surviving workers streamed back up the hills and across the desert to their mills.

It was the deadliest incident in world labor history. The Iquique Massacre, as it became known, broke the back of organized labor in the nitrate fields. Eventually the owners finally abandoned the *ficha* system, but at their own speed, in their own time. The Chilean military changed its rules so that soldiers would no longer be given active duty in their hometowns. Years later General Silva, the man who gave the order to fire, was assassinated by a relative of one of the Iquique dead.

THE WORKERS DID not know it, the owners did not know it, the government leaders did not know it, but the heyday of Chilean nitrates was nearing its end. In 1907, the year of the Iquique Massacre, two hundred mills were operating in the desert. By 1940 only a handful was still putting out nitrate. The rest were already ghost towns scattered across the desert. Today, Iquique, much of it looking as if it had been frozen in time in 1907, is working to reinvent itself as a resort town.

The *calicheros*, the *salitreros*, and the mill owners are gone, but not because the Chileans stripped the Atacama of nitrate, as the Peruvians stripped the guano. There is still plenty of nitrate in Chile.

What ended Chile's nitrate age was not shortage but plenty. Just as the workers were marching into Iquique, half a world away a German chemist was channeling his anger into perfecting a machine that would within thirty years bring down the entire Chilean industry. The man was Fritz Haber. The machine he was working on, it was said, could turn air into bread.

II

The PHILOSOPHER'S STONE

CHAPTER 6

I T WAS ALWAYS the same thing: Fritz Haber got excited and anxious and eager and intensely productive, and then his stomach tied in knots and he had trouble sleeping. He grew short-tempered. He called it nerves. About once a year the anxiety increased to a point where he could no longer focus effectively on science and had to take time off, spend a few weeks at a spa or a sanatorium, take a health cure. Afterward he would return, revived, ready to throw himself back into his research.

This time he knew it was going to be bad.

It had been building for months and came to a head at the 1907 annual meeting of the Bunsen Society for Applied Physical Chemistry in Hanover, Germany. Here, in front of an audience of Haber's peers, before the men he wanted most to impress, he was publicly insulted. To make it worse, the man attacking him was one of Germany's greatest researchers.

The insult came from Walther Nernst, a chemist about Haber's age who seemed to have achieved everything Haber had not. Nernst was a full professor of physical chemistry in the epicenter of German academic culture, the University of Berlin; Haber did research at a solid but unspectacular technical university in the south of Germany. Nernst was being spoken of as a future Nobelist for coming up with the third law of thermodynamics; Haber had not produced anything close to a great theory. Nernst was the protégé and heir apparent to Wilhelm Ostwald, Germany's reigning genius of physical chemistry; Haber had twice been turned down for a job in Ostwald's lab. Nernst was a Gentile; Haber was a Jew, a fact that might have played a role in Ostwald's rejection—Ostwald wrote that he turned Haber down because he seemed a bit too ambitious, a bit too pushy—code words, perhaps, for being a bit too Jewish. As if all that were not enough, Nernst had sold an idea for an electric lightbulb to a German firm for a million marks. He was a very rich man. Haber was not.

Their dispute seemed a minor matter, a question of a few data points related to the formation of ammonia, but it had festered and grown. Ammonia—a simple compound formed when a nitrogen atom combines with three hydrogen atoms—releases heat when it is formed. Nernst's laboratory had calculated precisely how much energy was involved in the reaction. But his numbers differed from some Haber had published back in 1905. Haber thought his numbers were right, Nernst thought his were, and neither man would back down.

Haber had done his original ammonia work for an Austrian company that wanted to find a profitable way to take nitrogen in the air (which cost nothing) and use it to make ammonia (which could be sold). Haber was hired as a consultant to explore the possibilities. He worked for several months on the project, which turned out to be quite difficult because the nitrogen in the air exists in a twinned form that chemists called N_2—two nitrogen atoms fused tightly together into a single molecule. The two atoms are so stable when linked to each other, so strongly tied, that they are almost impossible to separate. In this atmospheric form, called dinitrogen or, in chemical shorthand, N_2, they refuse all other relationships. The nitrogen bound in this form is inert, dead, unavailable to living things. Everyone breathes in and breathes out lungfuls of N_2 all day long and nothing happens. Early chemists, struck by the fact that candle flames went out and animals died when placed in a jar of nitrogen gas, gave it an early, now extinct name: azote, "without life." All the action in the air comes from another atomic twin, O_2, oxygen, a molecule eager to break up and enter into all kinds of reactions. Atmospheric oxygen makes things rust, makes things burn, and keeps you alive. Atmospheric nitrogen, it seemed, did nothing.

To turn N_2 into ammonia or anything else, its two nitrogen atoms first had to be torn apart. Once that was done the freed nitrogen atoms became exquisitely reactive, eager to bind to a number of other atoms to make an array of compounds, from fertilizers and explosives to the proteins and nucleic acids necessary to life. The problem Haber faced in his research for the Austrians was tearing apart the N_2, a process easier to imagine than accomplish. Chemical bonds come in a range of strengths. A typical chemical bond holding two atoms together, what chemists call a covalent bond, is strong. Sometimes atoms will be joined by two of these bonds at the same time—a double bond. That is much stronger. N_2 is held together with a triple bond, the strongest chemical bond in nature. Breaking it and freeing the individual N atoms requires enormous amounts of energy, heat on the order of 1,000°C, intense enough to to melt copper. The only thing in nature hot enough to break apart N_2 is a bolt of lightning.

Ammonia, by comparison, is fragile, which Haber found put him into a sort of scientific catch-22 while he was working for the Austrians: To tear apart the N_2 and release the single nitrogen atoms needed to make ammonia, the temperatures had to be so high that the system destroyed any ammonia that formed. Adding to his difficulty was the fact that the formation of ammonia—combining one of those freed nitrogen atoms with three hydrogen atoms—gave off quite a bit of heat of its own. It was almost impossible to save any ammonia before it burned up. Haber ran his experiments using special equipment made of platinum to withstand the heat (reaction chambers made of iron would have glowed cherry red at the required temperatures), succeeded in breaking apart the N_2, and tried to devise ways to quickly cool any ammonia that formed, but nothing worked well enough. He could never secure more than a trace of ammonia. A commercial process was an impossibility. He gave the company the bad news, wrote up his results, and published some of his data in 1905.

Nernst read Haber's report and thought that something was wrong. According to Nernst's theories, Haber should have found even less ammonia than he did. To check, Nernst set an assistant to work on the problem; the young man found, not surprisingly, that his boss Nernst was right. Nernst wrote Haber in the fall of 1906 that he had new, correct ammonia data that he planned to present at the coming spring's Bunsen meeting.

Haber's stomach started to clench. His reputation had been built on careful experimental work. To have Nernst question it publicly was unbearable.

Haber immediately started rechecking his data. One thing that Nernst had done differently was to use an air compressor to put the ammonia reaction under increased pressure, a trick for pushing it toward greater ammonia production. Haber added higher pressure to his experimental setup. A young chemist adept with machinery, Robert Le Rossignol, was brought on board to help figure out the apparatus. With Le Rossignol's help, Haber ran a new series of ammonia tests, heating the gases in quartz tubes to handle the increased pressure, and seeing what happened when different mixes of temperature and pressure were used.

The new results were closer to Nernst's but still significantly different. Haber kept getting more ammonia than Nernst thought he should. The yield was not high enough to interest any businesses, but the added pressure helped to squeeze out more ammonia, as did the use of different catalysts, substances that push forward chemical reactions without themselves being used up. By running the hot nitrogen and hydrogen gases over the right catalyst and putting it all under pressure, Haber could lower the temperature of the reaction, which helped reduce the destruction of the ammonia that formed and boosted his yields.

He unveiled his new numbers at the May 1907 Bunsen Society meeting. Nernst presented his own, lower numbers. Their "discussion" became heated. Nernst called Haber's numbers "highly erroneous" and noted drily what a pity it was that Haber's numbers were not accurate, because if they were they might hint at the possibility of a commercial way to make ammonia. He said, "I would like to suggest that Professor Haber now employ a method that is certain to produce truly precise values."

Those were fighting words. Haber, insulted, returned to his school and threw himself into even more ammonia research. Soon after, Haber's wife reported that her husband was suffering, his skin breaking out, and his stomach in knots.

Fritz Haber had been born into a thriving Jewish community in the town of Breslau, then located in the eastern part of Germany (it is now Wrocław, Poland). His mother died a week after giving birth. Haber was raised by his father, a successful dye and paint manufacturer who was rarely idle, rarely around the house, and, it seems, rarely satisfied with his son. Fritz grew into an impatient, insecure young man, devoted to proving himself, and eager to gain acceptance and achieve high position.

All of that, he felt, was within his grasp as a German Jew. Many Jews in Germany in the late nineteenth century felt fortunate. Many doors were open to them that were firmly shut in other nations: They were allowed to attend and even to teach at the great German universities; they could start successful businesses, as Haber's father had done; and they could practice professions like law, journalism, medicine, and the sciences. In Germany they excelled in all these areas. It was estimated, for instance, that up to 20 percent of Germany's scientists were Jewish. It was true that Jews were still barred from some careers—they could not join the German Officers Corps or serve in the highest levels of the Civil Service—and there certainly was an undercurrent of anti-Semitism, expressed openly at times, although rarely savagely (there were no pogroms in Germany). It was certainly true that poor Germans began to envy the success of Jews. Anti-Semitism was also especially deep, it seemed, among the old aristocracy of Germany, the Prussian nobles who controlled the military. It was even said that Germany's king, Kaiser Wilhelm II, a man whose policies seemed even-handed toward Jews, was a closet anti-Semite. Every German Jew, no matter how successful, had felt it. "For every German Jew," wrote Walter Rathenau (later Germany's foreign minister), "there is a painful moment that he remembers his entire life: the moment when he is first made fully conscious that he was born a second-class citizen. No ability and no achievement can free him from this."

Still there was a sense of hope among German Jews, a sense of pride, a sense that perhaps the great wheel of history was turning, that along with all the late-nineteenth-century excitement about worker's rights and women's rights there might also be a time when Jews would be integrated fully into German society. As the historian Amos Elon wrote, "In most other European countries, prejudice and discrimination seemed equally or more prevalent. With all its shortcomings, Germany stood out as a country where acculturation, social integration, and day-to-day tolerance seemed to have as good a chance as anywhere else in Western Europe." Many German Jews thought it was better even than France, where the Dreyfus affair had shown just how deep anti-Jewish feelings ran.

The important thing for Haber was not how Jewish he was,

but how German. This was true of many German Jews intent on leaving behind the old baggage of religious prejudice and concentrating on the birth of a new era. Intermarriage between German Jews and Christians became increasingly common during Haber's lifetime, rising from 8 percent of Jewish marriages in 1901 to almost 30 percent in 1915. It seemed at the turn of the century that the last bits of prejudice might wither away, and that Haber's dream of modern German Jews becoming, simply, modern Germans, would become reality. He, like many German Jews of the era, became a German superpatriot, a vocal supporter of the nation's ambitions, and a believer in Germany's shining future.

As a child he attended a progressive school—a simultaneous school, it was called—that mixed Jewish, Catholic, and Protestant students. It was a step toward assimilation, but it was not a complete success. He endured his share of taunts and teasing, and emerged from school with an unexplained scar on his face (it looked something like a dueling scar, ironically, a mark of honor among Prussian military men). But his dream of assimilation, his belief in Germany, was still untarnished. At the same time Haber's interest in chemistry blossomed, starting with semisecret chemical experiments in his room, where he mixed, heated, and watched things react until his father, unhappy with the smells coming from under his door, forbade it. When it came time to pick a profession, Haber proved a restless student, spending a few months here, a few years there, focusing mostly on chemistry but also writing a bit of poetry, working for his father for a time, and serving a required one-year stint in the military (he was allowed to join but not to serve as an officer). He learned how to carry himself with military bearing. He still planned to be a chemist but could not decide what kind he wanted to be. It was as if he were too smart, too curious. For years he flitted from school to school, in Berlin, Heidelberg, and Zurich, from mentor to mentor, from electrochemistry to physical chemistry to organic chemistry, constantly looking for acceptance and respect, constantly experiencing what he felt were rebuffs. He finally earned his doctorate in 1891 for studies related to the dye industry. He never seemed satisfied, was always critical of himself. "The thesis is miserable," he wrote afterward. "One and a half years of new substances prepared like a baker's bread rolls . . . lots of results that I cannot even publish because I fear that a competent chemist will find them and prove to me that the camel is missing its humps."

He earned money after graduation doing lab work in an alcohol distillery, a cellulose factory, an ammonia-soda factory, and a molasses plant. They all bored him. He researched furiously but without focus. He achieved little. The failure of this phase of Haber's career, one of his closest friends later wrote, was "total and protracted."

So he did what he thought was necessary. At age twenty-four he converted to Christianity. Haber had grown up in a Jewish household that was anything but orthodox—his sister remembered that Christmas trees and gifts were more important than Yom Kippur—so it seemed like a relatively small matter. He, like many forward-thinking German Jews, saw religion as as more formality than conviction. A baptismal certificate, the German Jewish poet and essayist Heinrich Heine wrote, was the "the entrance ticket to European culture." Conversion was certainly not uncommon among German Jews of the day. Between 1890 and 1910, about ten thousand German Jews converted out of a population of somewhere between a half million and a million, and the numbers kept increasing. (It is worth noting that a number of leading German Jewish scientists of the same generation, however, never converted, James Franck, for instance, Richard Willstätter, and Albert Einstein among them.) For many educated Jews like Haber, conversion was simply a way to remove a stumbling block. In the modern world, advancement would be based not on religion but on achievement.

Haber's god was not the god of Moses, not the god of Peter, but the god of Science. Pure rationality offered an escape from the old prejudices as well as a path to a better material life. Here again Germany was preeminent. Germany was the best place in the world to do science; German universities had the best professors; German industries had the most advanced facilities; Germans produced the most important discoveries and theories; Germans won the most Nobel Prizes.

So science became twinned in Haber's mind with national pride; the two things fused into a single way of life. Germany might be Europe's youngest major nation; it might be land poor, might lack the colonial holdings of other European powers, might be short of natural resources (other than iron and coal), might have poor soils and harsh winters, might be trapped between huge Russia to the east and the power of France and England to the west—but it was rich in scientific power. Belief in the nation, in the kaiser, in the future, in discipline, and in science would earn Germany its place in the world. Discipline had built the German army into the best in Europe. Discipline gave its government (under the "Iron Chancellor," Otto von Bismarck) its power. Science too was disciplined, rigorous in its methods and certain in its conclusions. Science had turned Germany's meager resources into great wealth, allowing the nation to make the world's best steel from its iron, the world's best machinery from its steel, and the world's best dyes and chemicals from its coal. These industries brought in money from around the world, making Germany rich enough to grow a vibrant, highly educated middle class—including many Jews. As long as Jews contributed to the spirit and accomplishments of science and the nation, Haber thought, they would be valued, tolerated, and eventually accepted.

HABER BURIED HIS Jewishness and threw himself into science. Slowly—too slowly for his ambition—he began to achieve success. He was hired at the University of Karlsruhe, a solid if unspectacular school on the Rhine south of Heidelberg, joining a chemistry faculty whose members might not be the most renowned in Germany but were well respected. Haber found that he was good at parties and enjoyed socializing, drinking, and talking about books as well as science. He loosened up a bit and made friends with some of the more liberal members of the faculty and local artists, creating "a bit of Bohemia in Karlsruhe," as one writer put it. He became known as a reciter of literature and teller of jokes and stories. He began to enjoy Karlsruhe, although the prominence of the school never matched the eminence of his own self-image.

Haber tamed his restless intellectual energies, settled down in the laboratory, and built a reputation as a precise experimentalist and an innovative thinker. He did studies on everything from the corrosion of pipes under city streets to the laws of thermodynamics, from electrochemistry and the loss of energy in engines to the reactions that lay at the luminous heart of flames. He wrote a successful book on the thermodynamics of gas reactions. He began to earn a solid reputation. In 1906 he achieved what was for any German a highly respected position: university professor (in Haber's case, of physical chemistry and electrochemistry). This meant not only a title of honor but also a secure, salaried position. Haber settled down, married a bright young woman—a Jewish girl from his hometown, the first woman to earn a Ph.D. in chemistry at her university—had a little son, and seemed destined for a long, productive, and relatively quiet career.

But he wanted more. He always wanted more. He wanted more money, more fame, more respect. He worked too hard, took on too much, and still suffered from nerves. His career was everything to him and he proved a distant and self-centered husband. His marriage suffered. "Fritz is so scattered, if I didn't bring to him his son every once in a while, he wouldn't even know that he was a father," his wife wrote. Then came the Bunsen meeting, the unbearable Walther Nernst, and the unforgivable insult of having his experimental data called "highly erroneous."

Not even a chemist as eminent as Nernst could be allowed to level such an accusation. Haber could not let it stand. He threw himself back into his research on ammonia.

CHAPTER 7

THE SHADOW OF Wilhelm Ostwald fell over everything that happened next, and the shadow of Ostwald, a giant of science, was substantial. Among his many achievements—energetic pioneer of the precise melding of physics and chemistry (and a founding father of the field called physical chemistry), able leader of his own research institute at Leipzig, ambitious founder of the German Electrochemical Society, and respected symbol of the close ties between German academic science and German industry—Ostwald was also one of the first to answer Sir William Crookes's call to solve the nitrogen problem. Back in 1900, seven years before Fritz Haber and Walther Nernst went at it in Hanover, Ostwald thought he had found the answer.

He came to the work not out of a desire to save humanity, but because he had been reading about the Boer War. Germans were very interested in the war, which pitted African farmers against British soldiers. The Boer farmers were generally of Dutch and German stock, and German public opinion was solidly against the British. It was, in a smaller way, a preview of the passions of World War I, and it started Ostwald thinking about what might happen if, God forbid, Germany ever got into a full-fledged war with England. His mind turned to the science involved, to the chemical sources of explosives and fertilizer. He knew that Germany depended on Chilean nitrate for both. He also knew that Britain had the world's best navy.

Putting the facts together led to a nightmare scenario in which a British naval blockade of Germany would cut the flow of nitrate from South America, cripple German farms, starve German citizens, and silence German guns. The British would force Germany to its knees. The only way to prevent this disaster was to find a way for Germany to make its own fertilizer and and its own gunpowder without relying on international trade. The answer, clearly, was somehow fixing nitrogen from the air. Following Crookes's challenge, a number of investigators around the world had started trying to find an answer. Most of the early efforts focused on machines that would burn the nitrogen out of the air, like lightning bolts did. The mechanical replacements for natural lightning were going to be high-energy electric arcs, and several teams in the United States and Norway were eagerly seeking ways to use electricity to trap artificial lightning bolts inside a box. This approach, however, had significant drawbacks—electricity was expensive and the process consumed huge amounts of it, the electric arcs burned out machinery and electrical connections, and the fixed nitrogen came out in the form of corrosive nitric acid—and Ostwald thought of a better approach.

Rather than burning nitrogen out of the air, he focused on a chemical method in which atmospheric nitrogen would be persuaded to combine with hydrogen gas to create ammonia. Essentially, he started to figure out almost everything that Haber did, only years earlier. Ostwald was an expert in the new field of catalysis, and he reasoned that the necessary reaction was going to involve a balance of heat, pressure, and catalyst. He did a series of calculations and then built a little test machine with a reaction chamber, a heater, input lines for nitrogen and hydrogen gases, and, for building up a bit of pressure, a bicycle pump. He tried various materials as a catalyst and found one he liked in the form of some common flower wire purchased at a local shop. When he put bits of the iron wire in the chamber and passed hot gases over it, he found in 1900 (just two years after Crookes's speech) that he could make an appreciable amount of ammonia. It required careful adjustment of temperature, pressure, and catalyst, and quick cooling for the ammonia, but when he got those factors right, ammonia appeared. It was somewhat surprising, and very welcome. Ostwald was growing unhappy at Leipzig, and a machine like this could be his ticket out. His former assistant Nernst had already gotten rich off of chemistry; now it was Ostwald's turn. He quickly applied for a patent and offered to sell his invention to chemical companies—for a million marks.

One of them, BASF (Badische Anilin- und Soda-Fabrik), then Germany's biggest chemical company, was very interested. Before an offer was made, however, BASF assigned a young chemist named Bosch, a novice who had been at the company for less than a year, to test Ostwald's machine. Everyone was unhappily surprised when Bosch's tests showed that Ostwald's apparent success was only that—apparent. The ammonia he thought he was producing from the atmosphere was actually the result of contaminants in his machine. It was a humiliating setback for Ostwald, who withdrew his patent application and pulled out of the race to solve the nitrogen problem.

He later compared the chemists' search for fixed nitrogen to the ancient quest for the philosopher's stone, the mythical substance that was said to turn lead into gold. Ostwald realized that finding an inexpensive method to fix atmospheric nitrogen would be even better, because it would turn air into gold. Both searches, however, carried risks. A number of medieval alchemists seeking the philosopher's stone went mad, were thrown into poverty, became obsessed by the search and corrupted by greed. The hunt for the stone could doom the hunter, as everyone who read Goethe's *Faust* understood. Ostwald had just a taste of it, so eager to sell his idea that he opened himself to humiliation by a stripling. Fixed nitrogen was the philosopher's stone of its time, and other seekers would experience their own obsessions, their own tragedies.

"MR. HABER IS a very busy, pushy man," Carl Engler wrote BASF in February 1908. It did not sound like a recommendation, but it was an attempt. Engler wrote as a senior chemist at the University of Karlsruhe, as a friend and fellow faculty member of Haber's, and as a member of the supervisory board of BASF. He was trying to help Haber get a consulting contract to help him further his nitrogen research. If Engler felt it necessary in his letter to be a bit critical of Haber, perhaps it was to emphasize that he was not unaware of Haber's "Jewish" traits, to show that he could be objective even though they were colleagues. "Personally, I have no interest whatsoever in BASF's obtaining Professor Haber's services," Engler wrote. He did, however, believe that the new project Professor Haber had been working on had great potential. He could vouch personally for Haber's "talents and energies" as a "thoroughly schooled expert" in electrochemistry, as well as "a sharp and clever dialectician." As for the money Haber was asking for his idea, well, "Because he is not unaware of his own worth and—just like the Ostwald school—would also like to make some money," Engler wrote, "he is of course not exactly inexpensive."

Money was what it was coming down to. The eight months between Nernst's insult at the Bunsen meeting and Engler's letter to BASF had been full of advances, excitement, and now the promise of great things—if only Haber could get a little more money. He and Le Rossignol had followed the line of pressure, raising it higher than Nernst had, much higher than Ostwald had, and getting better results when they did. The more they pushed the pressure, the lower they could drop the temperature. The lower the temperature, the more ammonia remained intact and the easier it was to harvest. The less the ammonia disintegrated, the higher the yield.

They also explored the effects of different catalysts under various conditions, varying three factors, pressure, temperature, and catalyst, and looking for the optimum balance. Through the summer and fall they had kept at it, refining and improving a tabletop machine in Haber's lab, a contraption pieced together from a high-pressure vessel, tubing, a heating system for the nitrogen, and a cooling system for the ammonia. Le Rossignol proved a wizard with machinery, designing improved fittings and new valves, inventing as he went. There was no single moment of breakthrough, just a number of small improvements and incremental advances. Finally, the increments began to add up to something. Thoughts of Nernst disappeared as they realized that it might be possible, with the right mix of conditions, to produce ammonia from the air in significant amounts—volumes great enough to attract the interest of the chemical industry. By the start of 1908 Haber and Le Rossignol were making ammonia from air at rates several times higher than anyone had ever achieved. It still was not quite good enough for industrial use, but it looked promising—so promising that Haber talked with Engler about the commercial prospects for his research. He thought he was on the trail of what might be a hugely profitable process, one that could conceivably turn air into ammonia in enormous quantities. To pursue it he needed better equipment, more assistants, money to turn Le Rossignol's designs into new devices.

Engler gave him an entrée into BASF. In particular, he figured that the head of BASF, Heinrich von Brunck, would be willing to gamble on Haber's ideas.

BRUNCK WAS A gambler and always had been a high-stakes gambler who never played with his own money. Not that he did not have plenty of his own money; he came from a family of means, owned a mansion surrounded by parklands, and enjoyed raising rare orchids in his greenhouse. He also earned a munificent salary from BASF. He was smart, far-seeing, and innovative. He had built BASF into a chemical powerhouse. His workers idolized him.

Brunck's first great gamble had involved the prized dye indigo. The German dye industry had become rich by making synthetic colors from coal to replace rare and expensive natural fabric dyes made from plants and animals. Germany had an abundance of coal, and German dye chemists were expert at transforming cheap coal—by heating it, distilling it, breaking it into its many component compounds, then refashioning these compounds—into expensive products. That was the beauty of the chemical industry. It could turn common ingredients into big money.

This was how BASF had started. For decades after its founding in the 1860s it was just another dye firm among many, all of them tinkering with molecules, looking for the next style-setting color, the next improvement in colorfastness, a bigger share of the world market. Competition was fierce in both the laboratory and the marketplace. To succeed, a firm had to not only maintain a steady stream of new discoveries, but also build the efficient worldwide marketing and sales operations.

While synthetic German colors were popular because of their novelty—almost all of them were new hues, colors rarely or never seen in nature, from methylene blue to Congo red—there were drawbacks. A number of the coal dyes faded relatively quickly from laundering or exposure to the sun. Despite the flood of new colors, millions of customers still preferred longer-lasting, better-known natural dyes, the most ancient and valuable of which was a beautiful, long-lasting blue called indigo. It was the holy grail of dyes. Lightly dyed with indigo, fabrics turned the azure of the sky. Deeply dyed, they approached a royal purple. Now we know it mostly as the blue of blue jeans, but its origins were much loftier. Indigo painstakingly gathered from Mediterranean snails had tinted the funeral garments of the pharaohs of Egypt and the ceremonial robes of the Caesars. In India a similar color made from a subtropical plant was used to dye rugs and saris (the word "indigo" is rooted in the dye's relation to India). As world commerce and trade grew after the Renaissance, indigo became one of the most important "spices" of the East India trade, filling the holds of European ships and making fortunes for traders. Indigo was the most prized, most valuable dye on earth.

Any coal-dye company that could find a way to make synthetic indigo would make a fortune. That was the prize Brunck wanted. He joined BASF as a young chemist in 1869 and, after working his way up the ladder to the post of technical director, threw the company into a search for synthetic indigo. It was his first great gamble. Several firms had already tried to make it in the laboratory, but none had succeeded. Brunck believed BASF could do it—if the company devoted enough scientists to the search. German dye firms had generally started as small operations, family firms inventing dye recipes through a sort of kitchen chemistry, a little of this, a little of that. Many of them had been slow to hire professional chemists. Brunck believed that the future belonged to those firms that switched to a larger-scale, more thoroughly scientific approach, with bigger, better-equipped laboratories full of highly trained professionals. That, he thought, was what indigo required. He persuaded BASF's board of directors to back his gamble. Their approval started a flow of money, which paid for cadres of chemists and allowed Brunck to devote entire sections of the company to the search.

He failed. After years of rising costs, synthetic indigo still eluded BASF and the board began to grumble. But as the 1880s turned into the 1890s Brunck saw something bigger emerging from the effort, something potentially more important than indigo. The infrastructure he had built for a specific dye was turning into something more: It was a way to transform a dye company into a modern chemical company. Large amounts of certain acids were needed for the indigo project, for example, so to save money his growing cadre of BASF chemists figured out better ways to make acids in bulk. They soon had excess acids to sell. Chlorine was another important raw material for dye processing; a BASF chemist found a better way to purify it and the company began selling it. All these processes took place in machines, and Brunck paid attention to the required engineering as well as chemistry, seeing the two as inextricable, pioneering the field of chemical engineering before people ever used the term. He saw BASF not as a miscellany of separate parts but as a single integrated machine, with each research team, each improved process, each new product feeding into others, making complex processes cheaper and more efficient. By the 1890s Brunck had gathered the world's largest staff of chemists (about 150 by 1899). The firm was making money from a variety of chemicals in addition to dyes, and Brunck made sure that the income was plowed back into more research for more new ways to make more products.

It was a great system, but it did not make everyone happy. Many of BASF's investors, for instance, would have preferred that company profits be paid in dividends instead of flowing in distressing amounts to research and development. Brunck knew better. The Industrial Revolution a hundred years earlier had been about steam and iron. The new one, the industrial revolution of the twentieth century, was going to be about R&D.

Germany's chemical industry led that change. BASF, for instance, was an early example of a multinational corporation, dependent on international sales, investing in companies in other nations (like the Norwegian arc process) with sales staffs salted around the world. It was also a progenitor of today's high-tech companies. As science moved forward, BASF had to move forward just as fast or risk losing its competitive edge. Chemistry at the end of the nineteenth century was racing at top speed. Discovery fueled discovery, every new finding, new theory, and new process informing and hastening the next. Chemists had become experts at manipulating molecules, changing old ones, making new ones, and every new molecule held the potential of huge profits but—and this was a key "but" that Brunck understood—only for a short time. Chemistry was unleashed; there was no holding back the pace of scientific discovery. Most leading chemists worked in universities, dedicated to open communication, quick publication, lectures at international meetings, and informal communication through letters. Nothing was a secret for long. Any breakthrough made in one laboratory would soon be duplicated or improved in another.

Discoveries made in industrial laboratories like Brunck's could be kept secret for a while, but even there, no matter what a company came up with—no matter how inventive BASF's chemists and engineers were—others would soon duplicate, extend, perhaps even steal their competitors' processes and products. The rise of the chemical industry was accompanied by a rise in industrial espionage. In a world where there were no secrets—at least not for long—only the innovative would survive.

That was the field in which Brunck and BASF were playing when they bet on indigo. As costs mounted through the 1880s and into the 1890s, some of the BASF directors lost heart and began calling for an end to the project. Brunck took them on, arguing the big vision, reminding them that if they gave up now, the money they had already spent would have to be written off as a loss. If they found a way to make synthetic indigo by the ton, the profits they made would make the money they spent now look trivial. They had to push forward, he argued, and they had to do it now. Their research to date put them ahead, but any slowing would lose them the race. BASF would do it, he said. They would find synthetic indigo. He built a network of allies on the board. He would not give up. When the indigo fight was over, Brunck had won, but he was no longer technical director. He was the head of BASF.

With a freer hand, he worked to perfect his business model. Huge machines and scientific talent were expensive and bankers were often slow to invest the needed sums (it took a lot of explaining to get a banker to understand dye chemistry), so Brunck used his company's growing income to build huge cash reserves—slush funds, some called them—so he could buy what he needed when he needed it, from a new railroad spur to a new factory, without second-guessing and delay. At the same time, his people perfected a marketing and sales organization, a product pipeline that got his goods sold quickly and in quantity before their short lifespan was over.

Innovation and discovery, research and development, fast-track marketing and sales, cutthroat global competition, cash stashed in hidden reserves: Brunck would have felt at home in any number of today's big corporations.

In 1897, after more than a decade and a half of research and an investment of eighteen million gold marks—an amount by some estimates equal to the total value of BASF—Brunck's chemists finally made synthetic indigo in bulk. The company ramped it up, making it by the ton, and quickly put it on the market. He was right. It was an immediate and enormous success. It was better than natural indigo in the eyes of many users, its quality more consistent, and its selling price cheaper. It quickly replaced natural indigo. Profits soared. Within three years, thanks to indigo, BASF claimed its place as the world's number one chemical company.

Brunck, however, was already looking ahead. Synthetic indigo was the last great hurrah of a dying (pun intended) industry. The secrets of the German dye industry were being copied around the world, competition was heating up, and the market was being saturated with a rainbow of competing colors. If nothing was found to replace dyes, German firms would start tearing each other apart for smaller pieces of a shrinking market.

It was time for something new. Something very big, big enough to replace dyes. Something that would make money in undreamed of quantities. Brunck surveyed the field. Bayer and Hoechst, two of BASF's major competitors, were looking for pharmaceuticals (Bayer had made a fortune off of aspirin, a pain reliever made from coal). Other firms were exploring synthetic fibers and photographic chemicals; BASF did a bit of that itself.

Years before Crookes gave his speech, Brunck was already thinking about fixed nitrogen. Germany depended on Chilean nitrates for its fertilizer and gunpowder, buying it by the ton and shipping it by the fleet halfway around the world. It was a dangerous addiction. Brunck, like Ostwald, realized that any interruption in the trade would mean disaster for the German economy. That was the attraction of the air. Tons of nitrogen lay quietly over every acre of land, just waiting to be made into fertilizer—if some way were found to extract it at prices lower than the Chilean import. By some estimates, the worldwide synthetic nitrogen market was two or three times bigger than the entire global dye market—some researchers said ten times bigger. Brunck wanted his firm to be first in.

The issue was cost of production. Fertilizers and explosives were made and sold by the tens of tons. The difference of a few pennies per kilo in production prices meant the difference between profit and loss. Chemists not only had to find a way to fix atmospheric nitrogen, but they had to find a cheap way if they were going to compete with the Chileans. If and when that method was found, the investment in finding it would be well repaid.

Brunck started working on the arc process two years before

Crookes delivered his address. In 1897 BASF assigned a chemist named Otto Schönherr and an engineer named Johannes Hessberger to the task of making effective electric arcs and finding better ways to flow air around them in a furnace that would allow them to capture the nitrogen burned out of the air. The two men beavered away for years, stretching their crackling arcs to great lengths and finding an innovative way to twist air around them. It took a decade and a substantial investment to get BASF's Schönherr furnaces to the point of a full industrial tryout in Norway, where BASF had invested heavily in Norsk Hydro, a company that had purchased a number of waterfalls and was developing another arc method. The arcs were not perfect, electricity was still too expensive, the products from Schönherr furnaces cost more than Chilean nitrates, and Brunck was not satisfied, but at least BASF would have something in place when, as Crookes predicted, the Chilean desert ran dry.

This was not a perfect solution. Then BASF received Haber's letter, describing his new advances in making ammonia.

CHAPTER 8

FRITZ HABER WROTE BASF a six-page review of his nitrogen-fixing research, including both the arc process and his ammonia work. He reviewed the latest experiments he and Robert Le Rossignol had conducted, emphasized their constantly improving results as they refined their equipment and worked at higher and higher pressures, and noted that his system was now good enough to more than hint at the possibility of commercial value. Haber also mentioned that a rival firm, Hoechst, had already expressed interest in his ideas. He knew how business worked.

On March 6, 1908, two contracts were signed between Haber and BASF, the first for further studies of the arc process, the second for high-pressure ammonia research. The company got what it wanted: ownership of any resulting patents and Haber's agreement not to publish any of his findings without permission from BASF. Haber got 10 percent of any net earnings (determined through an intricate set of calculations) that BASF made from processes he developed. It was clear to Haber that the company was most interested in arc work and funded his ammonia research "out of personal consideration for my wishes, and not out of confidence in the matter itself," Haber said.

It was a good deal for Haber. He and Le Rossignol needed expensive new equipment to continue their studies—bigger compressors, better instruments, stronger fittings—and BASF would buy it for them. The deal included money for more assistants. Perhaps most important, Haber would be able to keep his job at the university while he did the research. The BASF money, meanwhile, would more than double his salary.

Once Haber and Le Rossignol had the funding they needed to explore the new world of high-pressure chemistry, their work moved forward quickly. Nernst had the right idea, using pressure to push the reaction toward ammonia, but had not used enough. Haber and Le Rossignol boosted it to levels earlier chemists would not have dared because their equipment would have exploded. They ran experiments between one hundred and two hundred atmospheres (the number indicates how many times higher than normal atmospheric pressure), the same pressures found about a mile beneath the ocean, pressure enough to crush most modern submarines. It was easily enough to burst most metal containers. Le Rossignol figured out how to harness it, running the reaction in a thick-walled quartz tube encased in an iron jacket, with newly designed valves and fittings that would hold. The higher they pushed the pressure, the more ammonia they made. The worry was only that at those pressures it would be difficult to scale up any process to industrial levels; it would be difficult to design a reactor strong enough to stand it. That, however, was someone else's worry, not Haber's.

Two months after Haber signed his deal with BASF, he heard that Walther Nernst too was working on his own solution to the ammonia problem, and was getting close. It was just talk, as it turned out, but it kept Haber focused. He threw himself into his project, pushing the pressures to the limit of the compressor.

Then he started improving every other part of his machine, refining the system for getting the ammonia away from the reaction as fast as possible and cooling it; devising ways to handle the input gases, the nitrogen and hydrogen, as efficiently as possible. Haber's laboratory came up with a sort of heat-pump idea, using the heat released by the reaction (freed nitrogen combining with hydrogen to make ammonia) to warm the cold incoming hydrogen and nitrogen, recirculating the hot unreacted gases and running them back through. Le Rossignol and Haber designed and redesigned the gas preheating and circulation systems through 1908, eventually patenting their ideas.

Increasing the pressure allowed them to lower the temperatures, which left more ammonia intact. Now, instead of working at 1,000°C, they were running their tests at temperatures as low as 600°C without decreasing the yield. They were making incremental progress, balancing heat and pressure, making the system more efficient, getting a little more ammonia each time. But it was still not enough. The yields were still too small for a commercial process.

They focused on the last variable, the catalyst. They, and Nernst too in his work, had used iron, the same catalyst that had proved fatal for Ostwald. As it turned out, Ostwald had the right idea, but an insufficient setup—the iron worked well only when mixed with other substances at increased temperatures and pressures. Haber wanted something better. He tried powdered nickel, manganese, and platinum; none did the trick. So he turned to more exotic elements. One of the many consulting jobs Haber had done involved looking for ways to make better filaments for electric lights, and the company he had worked with had supplied him with a number of rare elements. He still had samples of many of them. He tried them in his ammonia machine.

In March 1909, one year after signing with BASF, Haber had a breakthrough. His team put a bit of one of his lightbulb elements, a bluish black brittle metal called osmium, in the high-pressure chamber, heated it, and ran hot nitrogen and hydrogen over it. The yield of ammonia shot up. The excitement was tempered, the machine checked and readjusted, and the osmium retested. The yield was again high—much higher than they had ever seen—high enough to work industrially, high enough to supply the world with fertilizer. No one knew why the osmium worked (no one knew much about osmium at all), but it worked. Haber ran upstairs from his laboratory and went lab to lab down the hallways, poking in his head and calling out, "Come down. . . . You have to see how the liquid ammonia is running out!" Those who came watched together as drops of chilled liquid ammonia collected in a flask. "I can still see it," remembered one of them decades later. "There was about a cubic centimeter of ammonia. . . . It was fantastic." One cubic centimeter is a little less than a quarter of a teaspoon.

It was enough. Haber's experimental machine was small, the size of a few table lamps sitting side by side on a lab bench. It would simply have to be scaled up. A small yield from a small machine would mean a big yield from a big machine. When the big machines started working, the drops of ammonia would turn into streams, then rivers. On March 23 Haber reported his success to his partners at BASF, with the recommendation that the firm purchase all the osmium it could find.

The exciting news traveled quickly to the top, kicking off a high-level argument. August Bernthsen, BASF's director of research, was dubious. He could not believe BASF was even considering developing a process that required pressures in excess of one hundred atmospheres. No container he knew of could withstand that kind of pressure. The apparatus would explode. Haber had been able to do it only because his little tabletop demonstrator used reaction chambers drilled out of solid quartz. BASF could not produce ammonia drop by drop as Haber had; the firm had to make it by the ton and no quartz crystal on earth was big enough for an industrial-sized machine. A device that worked at very high temperatures and extraordinarily high pressures, without stopping and without breaking? Nothing like it had ever been considered, much less built. It almost went without saying that the rarity and cost of osmium was also a factor. No one knew how fast the osmium would be contaminated in the machine and lose its catalytic power or how often it would have to be replaced. The entire world supply of osmium amounted to, what? no more than a few hundred pounds? Bernthsen told Haber that the firm had no interest.

Carl Engler once again stepped in. Engler wrote Brunck directly, making Haber's case for him. Haber met with three of the top men at BASF who came to Karlsruhe to examine his machine personally. Heinrich von Brunck himself showed up, along with the Bernthsen and the chemist who had shot down the Ostwald machine, Carl Bosch. They looked at Haber's device and started talking. The discussion quickly turned to Bernthsen's concern about high pressures. Haber confirmed that at least one hundred atmospheres would be necessary to make his system work (this turned out to be an understatement; the machine worked best at two hundred atmospheres). "One hundred atmospheres!" exclaimed Bernthsen. "Just yesterday an autoclave at seven atmospheres exploded on us!" No metal cylinder on earth, he thought, was strong enough to contain Haber's little inferno, with its temperatures hot enough to temper steel and its pressures greater than the surface of Neptune. He thought Haber's machine was impossible.

Brunck turned to Bosch, who had been working on nitrogen chemistry at BASF for the past eight years, ever since the Ostwald incident. Brunck asked him what he thought.

Bosch thought carefully before he spoke.

BOSCH KNEW SOMETHING the rest of the men in the room did not. He knew about metals. Bosch had grown up around tools; his father was a successful gas-and-plumbing supplier in Cologne who had fitted their home with a complete workshop and given his boys free run. As a boy, Carl once took a carpenter's plane to some of his parent's bedroom furniture (he wanted to see how it looked beneath the surface) and disassembled his mother's sewing machine. He had an open door into his father's business and visited the workmen often, learning about pipe fitting, soldering, machining, and woodwork. Bosch had though seriously about a career in metallurgy. As a young man he did an internship at a metalworking firm, took a summer job in a blast-furnace plant, and studied mechanical engineering for two years before switching to chemistry.

Bosch was hired as a chemist at BASF fresh out of university in 1899. From the start, he was a bit of an oddity. Chemists were kings at BASF, highly educated and highly prized experts responsible for discovering the new products that kept the firm afloat. Many of them came from the best universities in Germany. Most wore suits, ties, and stiff collars under their lab coats, discussed art and music and literature in their off-hours, and maintained a clear dividing line between themselves and the firm's common laborers. Dinner with executives and the scientific staff above you—sometimes supplemented by the wooing of a director's daughter or niece—was the proven route to advancement.

Bosch did none of these things. He often took off his jacket once he got to work, loosened his tie, picked up a hammer or wrench, and started banging away on some machine. He kept to himself mostly and seemed uncomfortable socializing, except on the weekends when he liked to drink beer and go bowling. He seemed unafraid of blisters and stains. It did not augur well for his future. Early in his career Bosch was found by a BASF executive in a plant workroom, his sleeves rolled up, face gleaming, stirring something in a vat. "My dear man," the executive said, "if you think such foolishness will help you rise to the top at BASF, you are sorely mistaken."

Then came the incident with Wilhelm Ostwald. Bosch had been at the firm less than a year, working mostly on dyes, when he was pulled out of the ranks of junior chemists and given a special job: replicating Ostwald's machine for making ammonia. It was a signature honor for the young man, conferred presumably because it was known that he was adept with machines. It did not take him long to make one in the BASF labs. The only problem was that he could not get it to work. He told his supervisor, who told Brunck, and word reached Ostwald himself, who was, to put it mildly, surprised by the news. His machine had worked in his laboratory and produced appreciable yields of ammonia (although not as high as Haber later got). He assumed that the young chemist BASF had put on the job was doing something wrong. He made sure that Bosch was supplied with samples of his personal flower wire, the catalyst he had used in his original machine. When Bosch put it into his BASF reactor, heated and flowed in the gases, and pumped up the pressure with his bicycle pump, he finally—to his great relief—made some ammonia.

Then it stopped.

It kept happening. Using Ostwald's wire, the apparatus would spit out a little ammonia, less than Ostwald had gotten, but at least it worked. Then it would stop. A fresh dose of Ostwald's wire gave a bit more, and then it stopped. Perplexed, Bosch went to the library and looked up everything he could find on reactions between nitrogen and iron, hydrogen and iron, and ammonia and iron, and after digging, he found his answer. Heating ammonia with iron, he found, could produce a substance called iron nitride. Iron nitride, when heated in the presence of hydrogen, could form ammonia. In other words, if the system was contaminated with iron nitride, the results would be thrown off. Bosch tested Ostwald's flower wire for iron nitride—and found it. It looked as if the great chemist's magic flower wire had been heated in the presence of ammonia and become contaminated with iron nitride; the nitride on the wire, heated with hydrogen gas, formed ammonia. Ostwald, in other words, had been making ammonia not from the air but from other ammonia. Uncontaminated iron did nothing. Ostwald's machine did not work.

Bosch carefully chronicled his findings and dutifully reported them to his boss, who reported them to Brunck. It was almost unbelievable that Ostwald could have made such an error, and Bosch's work was carefully checked. The young man was right. The news was sent to Ostwald, who had been looking forward to making a fortune from his discovery. He was outraged. He had already applied for a patent. He could not believe it, and wounded pride showed in his reaction. "When you entrust a task to a newly hired, inexperienced, know-nothing chemist," he hastily wrote BASF, "then naturally nothing will come of it." He stormed into BASF to check it himself, and there followed what Bosch remembered as "rather heated discussions." When the arguments were over, the facts remained: Bosch was right. It looked as if someone in Ostwald's lab had used the same flower wire both to study the formation of ammonia and its decomposition back into its constituent nitrogen and hydrogen, and his catalyst had been contaminated with iron nitride. Ostwald quietly withdrew his patent application.

THE BASF MANAGEMENT liked what they had seen in Bosch, the young man's willingness to take on a giant of chemistry, his ability to find answers, the backbone he showed when Ostwald challenged his results. He was soon put in charge of all nitrogen research at BASF (which included the Schönherr furnace, the deal with Norsk Hydro, and Bosch's own ongoing efforts). Brunck himself kept an eye on the young man, tracking the progress of the nitrogen work and learning to appreciate Bosch's strengths, his appreciation for the engineering side of the enterprise, his careful, quiet, considered opinions. Bosch was not apt to rush into anything.

Now, nine years after the Ostwald incident, Bosch sat at the table and listened to Haber answering questions about his ammonia process, so similar to Ostwald's. Bosch thought Haber got it, the right mix of catalyst, temperature, and very high pressure. The osmium was the biggest problem; there simply was not enough of it in the world to make ammonia at the volumes BASF wanted. That would have to be addressed. But Bernthsen was right: the big drawback was the pressure. It was one thing to make ammonia in a little tabletop machine and quite another to build industrial-sized reactors capable of holding that pressure. Nothing like it had ever been done.

All that was only part of the equation. Bosch knew how much was riding on the nitrogen project. He had been working for the last seven or eight years to find the breakthrough that Brunck and the company needed, and had come up with nothing. The best hope BASF had was still the Schönherr furnace, the project that had already been going for years when Bosch was hired, the arc machine that depended on deals with the Norwegians for cheap electricity. The Norwegians had their own version of an arc machine that they liked better. The Schönherr furnace would never make BASF the money it needed. Even if it proved a success, Bosch would never benefit much because he had done little to develop it. He needed a win. He had married after being put in charge of nitrogen at BASF and he and his wife were having children. He needed something that would carry his stamp, something that would ensure his future. If the Schönherr furnace was all the company ever found, Bosch would remain a middle manager forever.

That left the Haber machine as everyone's best chance. But it was a long shot. It was all unexplored territory. The system would require huge compressors—the biggest in the world— new fittings, new valves, new measuring devices, a whole new technology. It was not just the heat, the pressure, the catalyst, but all the ancillary machinery, the heat recirculators, the ammonia cooling, not to mention making the pure nitrogen and hydrogen input gases in quantities never before achieved.... It would cost millions to turn Haber's little experimental device into a full-sized industrial machine. The chances were excellent that it would fail. When that happened, Bosch's career would be over.

Bosch looked at Haber's machine. Then said to his bosses, "I think it can work. I know exactly what the steel industry can do. We should risk it."

BRUNCK WAS A gambler but a careful one. He immediately boosted Haber's research support but held off making any other commitments until a few things were settled, notably the discovery of a replacement for osmium. BASF was already in the process of cornering the world osmium market, but even if the company did, there was simply not enough of it to work. All the osmium on earth, devoted to making ammonia, would never make enough. A more common catalyst would have to be found. Haber and Le Rossignol quickly found one in the form of another relatively uncommon element—uranium—which provided results almost as good as osmium and was available in the amounts needed.

The discovery of a more common catalyst was, Bosch remembered later, "A particularly encouraging and for us decisive fact." But there was one final step before BASF marshaled its forces behind Haber's process: The machine had to be seen in full operation over a substantial period of time, producing ammonia continuously, without breaking down. From an industrial standpoint, reliable, continuous operation was a key just as important as the catalyst or any other factor. Haber's machine was designed to make ammonia not in batches, like many chemicals were, but in a flow, a continuous flow that would have to be maintained, uninterrupted, for long periods of time to be profitable. Bosch arranged a trip in early July 1909 to Haber's lab in Karlsruhe for himself, BASF chemist Alwin Mittasch, and one of the company's mechanical engineers.

They arrived to find Haber's laboratory in an uproar. One of his people, while tightening a connection on the ammonia machine to get it ready for the visitors, had applied a bit too much pressure. A seal had started leaking and the pressure could not be raised to the proper level. Until it was fixed, no demonstration was possible. The Germans have a name for the way an inanimate object like a machine can pick the worst possible moment to break down: *Tücke des Objekts,* "the spite of things." There was nothing to do but fix it. As Bosch and his co-workers waited, the repairs dragged on. Finally, Bosch began glancing at his watch. He had another appointment, he told Haber, and unfortunately would have to leave very soon. Haber could see a disaster in the making. This would underscore what Bernthsen had been talking about: Any machine designed to work at such high temperatures and pressures would fail. Bosch finally left. Mittasch and the technician stayed behind in hopes they might still see something worthwhile.

Haber might have imagined his future going out the door along with Bosch. But he and Le Rossignol kept at it. Finally the repair was done, the heaters and pumps warmed up, the gases began to circulate, the critical pressures and temperatures were reached, and at long last, ammonia began to flow. According to one version of the story, Mittasch pressed Haber's hand as the product began gathering in a small flask. They all watched, chatted about chemistry, asked questions, and checked the machine as the ammonia collected, drop by drop, about a cup every two hours. The men calculated the yield. It appeared that between 6 and 8 percent of the nitrogen that went in one end as N_2 gas was coming out the other as fixed nitrogen in the form of ammonia—a yield high enough for commercial development. After five hours of flawless performance they turned it off and left to make their report.

Mittasch returned to BASF completely convinced that

Haber had done it. His machine was capable of changing atmospheric nitrogen into ammonia in substantial amounts. Even more important, thanks to Haber's gas- and heat-recirculation systems, the machine did it economically, consuming just a fraction of the energy that the arc process did. If it could be ramped up successfully, if much bigger reactors could be built, it would be possible to produce fixed nitrogen in bulk at a cost competitive with the Chileans. There was no end to the nitrogen in the air, so in theory there was no end to the amount of ammonia they could produce. Germany would no longer have to rely on shipping its raw material for fertilizer and gunpowder halfway around the world. BASF would dominate a growing world market. And Crookes's challenge would be answered. Humans would be able to make as much fertilizer as they needed, forever, from the air. Global starvation would never again become a critical issue.

The demonstration was a small machine producing a small amount of ammonia for a small group of men. But it marked a turning point in human history. One historian later compared those hours in Haber's lab to the Wright brothers' flight at Kittyhawk, or Edison's discovery of a successful light bulb.

Haber was ebullient. With Mittasch's support, he felt, BASF would now back his process and build it into an industry. He would share in the profits. He had not only answered Crookes's challenge, but he had beaten Nernst and succeeded where even Ostwald had failed. He had found the answer to one of the biggest scientific challenges in the world. He was about to become famous. He was about to become rich.

He had found the philosopher's stone.

CHAPTER 9

FRITZ HABER THREW a party and everyone got drunk. All the members of his laboratory were there, and many of his university colleagues, and friends, and anyone that anybody wanted to bring along. When it was over, one partygoer said, "We could only walk in a straight line by following the streetcar tracks."

After everyone sobered up, things started to move quickly. Carl Bosch was eager to move ahead, to wash the taste of years of failure from his mouth; he wanted to start building a bigger prototype machine as quickly as he could at the BASF laboratories in Ludwigshafen. He and Haber worked together closely for the next few weeks, exchanging visits, writing letters, conferring about details of design and operation. Bosch wanted to know everything Haber knew. It was not long before Haber, complaining of stomach pains and exhaustion, retreated to a Swiss health resort.

Bosch pushed the project forward. He started snapping up talent, creating the core of a fast-growing research group devoted to enlarging and improving the Haber machine. Alwin Mittasch was put in charge of catalysts. A promising young engineer, Franz Lappe, was recruited to guide machine design. BASF wanted everything new to be patented, and much of what they were planning to do was going to be novel. People were hired to help with the paperwork, and ties were established to the sections of the company that handled legal matters. Bosch worked with Haber, crafting patents as soon as the chemist returned from his health resort. They started with a general approach to the process and added supplementary patents, such as one for the way heat was captured and recycled, all done under BASF supervision. Before it was over, there would be dozens of patents, covering every new aspect of the process.

Word of Haber's success began to spread. Other firms began nosing around, looking for information, trying to lure Haber and Robert Le Rossignol away. Le Rossignol snagged a lucrative position at a chemical and electrical firm in Berlin. Haber started receiving letters from the top chemical and electrical firms in Germany. His stock was rapidly rising. He had not been happy with the complicated revenue-sharing agreement he had made originally with BASF and now felt that he was in a good position to renegotiate.

In October, three months after his successful demonstration, Haber told BASF that he had been approached by Auer (Deutsche Gasglühlicht-Anstalt), a big Berlin firm that was something like the General Electric of Germany. Auer's head, the fabulously wealthy Leopold Koppel, made Haber an astronomical offer—more than eight times his current annual income, plus a complete state-of-the-art laboratory, plus the position of chairman of the supervisory board—if, among other research efforts, he would help them develop a nitrogen program. Of course, Haber told BASF, he was tempted by this generous offer. Who would not be? In light of this offer he hoped BASF might revisit his contract.

It was clear that Haber wanted a new deal. There followed the expected expressions of surprise and dismay from BASF about Haber's concerns, and affirmations of loyalty from Haber. Talks were started. Eighteen months after Haber first signed with BASF, his original agreements were replaced with a new ten-page contract giving Haber a guaranteed annual income of twenty-three thousand marks (including eight thousand marks for equipment and assistants), plus a few pennies on every kilo of ammonia the company produced. The new arrangement greatly favored Haber. He was not only getting a guaranteed base salary, but if things went as everyone hoped, if BASF made tens of thousands of tons of ammonia per year, Haber's pennies per kilo would make him a very rich man.

BASF was not thrilled at having to renegotiate, but the company could not risk Haber's departure. Its corporate displeasure was written into the new contract, where BASF noted that it "resented" Haber's "suspicions" about the original royalty arrangement and concluded, "In future we would want to set certain limits to Mr. Haber's somewhat unrestrained ideas."

DESPITE THE CORPORATE displeasure, Heinrich von Brunck made sure that the nitrogen team had everything they needed. The head of BASF believed that the complicated machinery was going to work, that the delicately balanced chemistry was going to be tamed, and that synthetic ammonia from the Haber process was going to be his next big breakthrough, the next synthetic indigo, the key to keeping his company well ahead of the competition. He gave Bosch what amounted to a blank check. Whatever equipment Bosch needed would be provided. Whoever Bosch wanted to work on the project would be hired. Whenever Bosch needed something internally, personnel or administrative support, his way would be smoothed.

Brunck trusted Bosch, and the more they worked together the closer the two men became. Soon others at the company began to understand that the quiet, practical Bosch had become Brunck's protégé. They had similar administrative styles—they were both gambling on nitrogen; they both appreciated the importance of machinery along with chemistry; they both liked to hire talented people and give them a relatively free rein; they both disliked yes-men—and as they began to talk more they found that they both had the same idea about the chemical industry. They both understood that the game could be won only through ceaseless, rapid-fire innovation, with the next great idea going into development as soon as the previous one went on the market. They both understood the need for speed and efficiency in order to make big ideas into reality. They both thought a decade or two down the road.

They were both, in short, technological optimists. Problems would be solved. Solutions would be found. Their optimism was not shared by Haber, who, despite his enthusiasm while selling his ideas to BASF, had doubts about scaling up his ideas the way Bosch planned. At one point he himself had considered building a bigger machine but had given up the idea. Most of the ammonia problems that he and Le Rossignol had overcome had been solved in part because their machine was small. The reaction chamber was small enough to carve out of quartz. The compressor they needed was available because it only had to squeeze gases into a small space. The rare catalyst they discovered was possible because they needed only a small amount. As the size of the machine went up, every problem grew exponentially: bigger compressors than anyone had ever built; stronger and more complicated valves; reactor chambers that would have to be made out of some new material; high-pressure casings and fittings that could hold the pressure; as-yet-undesigned meters for tracking gases, pressures, and flows. Then there was the problem of making pure hydrogen and nitrogen gases— input for the process—in quantities never before produced.

Bosch bulled ahead, devoting himself to the sole goal of transforming Haber's two-and-a-half-foot-high laboratory machine into something the size of a factory. He knew about the potential difficulties, but he also knew that if he started thinking too hard about all of them all at once he would be over-whelmed. So he stopped thinking and started working. He wanted to build a prototype machine ten times the size of Haber's as quickly as he could. The first thing he did was outfit a special machine shop to make the many parts they would need. He started hiring people to help Lappe and started them thinking about valves and gauges and the quality of steel for the reactors. He gave Mittasch money to expand his catalyst testing laboratory. Bosch liked machines. He organized his people as if he were building one. From the start he proved himself an able administrator and a master coordinator.

Equipment design and catalyst development ran on parallel tracks. Haber had found only two elements, osmium and uranium, that worked well in his machine. BASF by now had purchased all the osmium in the world—about one hundred kilograms at a cost of four hundred thousand marks—but even in a best-case scenario, using all that osmium as the catalyst, the company could produce only about 750 tons of ammonia per year. BASF intended to make ten or twenty times that much. Uranium was the backup, but that too was relatively rare and expensive, and—to BASF's dismay—it was found that uranium lost much of its catalytic effectiveness after exposure to air or water. Bosch told Mittasch to find another catalyst.

The search was a hit-and-miss affair. Earlier catalysts used for other processes had been found among metals like iron and platinum, but when it came to ammonia Haber's work had shown that sometimes the unlikely candidates like osmium were the best. Mittasch started testing everything. The perfect catalyst would not only speed the reaction; it also had to be stable under high temperatures and pressures over a long period of time. That meant each catalyst test had to run for days. The only way to screen effectively would be to run a number of tests at the same time, in a number of small test machines. The engineers started working on them. Within weeks the BASF teams came up with a design for a stubby machine, about a foot high, electrically heated and air cooled. Best of all from Mittasch's standpoint, the catalysts were loaded, two grams at a time, into a cartridge that could be easily inserted into and removed from the test machine. Mittasch ran more than twenty of the machines at the same time, testing catalysts simultaneously, day and night, one after the other: platinum, palladium, iridium, ruthenium, element after element. Nothing he found worked as well as osmium. Then again, no one was expecting quick results.

It came as a pleasant surprise when in mid-September 1909, just two months after the decisive demonstration at Karlsruhe, Mittasch found something interesting. Bosch had remained keen on iron despite Wilhelm Ostwald's failure, so along with all his more exotic elements Mittasch had tested pure, simple iron. It did nothing. He also tested a battery of iron-containing minerals. This was where things began to look interesting. One of them, a Scandinavian magnetite (a naturally occuring combination of iron and oxygen), unexpectedly produced a good amount of ammonia. Other magnetites were tested with disappointing results. But this one particular sample, from a mine in northern Sweden, worked like magic.

Bosch ordered follow-up studies. Natural minerals had their own personalities; rocks with the same scientific names and general chemical makeup could differ in subtle ways depending on the mine from which they were dug. There were impurities and eccentricities. Mittasch figured there must be something else in this Swedish magnetite, a trace amount of some other element, a contaminant of some sort that made it work. To pick the problem apart he started testing pure iron mixed with known amounts additional elements, from aluminum to ytterbium, first adding one additional element at a time, then adding them in couples and groups and in various molecular forms and in varying proportions, looking for the perfect mix. He called these added materials promoters. Gradually, as his tests continued, a picture began to emerge. Iron made a solid base for the catalyst but did nothing until a promoter was added. The right promoter somehow switched the iron on, made it work. The trick was going to be finding exactly the right promoters, in the right proportions. He kept testing.

The work was slowed only by occasional mishaps when a test machine cracked under the pressure and exploded, spraying the laboratory with metal shards and bits of catalyst. Even the accidents were useful. The equipment designers removed the broken machines, autopsied them, and used what they found to make better models. An effective system began to emerge, with Mittasch's catalyst work performing double duty as a testing arena for machine design, for trying out ideas for valves, pumps, gauges, fittings, and heating systems and for finding better ways to contain the pressure and improve performance.

Bosch made a bunker of the catalyst test lab, putting up protective walls, armoring the test ovens in metal, and jacketing them in concrete. If they exploded, they exploded. The work could not be allowed to stop because of a little shrapnel. Mittasch kept testing.

He gradually zeroed in on a usable catalyst. Adding aluminum oxide to the iron base gave results almost as good as Haber's osmium. Adding calcium to the mix made it even better. It was not quite ideal, but it was close. The new catalyst could be damaged by sulfur or chlorine or any number of other contaminants that might get in with the input gases, but apart from that it was good: solid, cheap, stable, easily made, easily transported, and effective. It was an improvement in almost every way over osmium. Mittasch's careful experiments opened a new era in catalytic chemistry, with an emphasis on promoters rather than pure elements.

In January 1910 BASF told Haber about the new catalyst. It came as a surprise. "I am extremely happy that Dr. Bosch and his assistants have succeeded in . . . making such a great advance," he wrote the company, "and I congratulate him and you. But it is remarkable how in the course of things new special features always come to light. Here iron, with which Ostwald first worked and which we then tested hundreds of times in its pure state, is now found to function when impure. It strikes me again how one should follow every track to its end."

It marked the end of Haber's importance to the company. BASF now had his patents, knew what it needed to know, had already started making significant advances on its own, and no longer needed him. With the discovery of this better, cheaper catalyst, the ammonia machine slipped out of Haber's grasp. Communication between Bosch and Haber started to dwindle.

Haber did not much mind. Income assured, he turned his attention to his reputation. He wanted to let the scientific world know about his achievement. BASF, of course, was interested in publicizing as little as possible about the ammonia process, keeping it proprietary for as long as possible. Letters were written back and forth through the late winter of 1909 to 1910 until finally it was agreed that Haber could announce the general outlines of his discovery as long as he kept quiet about the operational details. On March 18, 1910, he made the first public announcement of the discovery in a speech, "Making Nitrogen Usable," delivered to a group of scientists at his university. He took the Crookes approach, talking about the need for fertilizer, the extent of nitrate imports from Chile, and the fixation of nitrogen by plants. He talked about the need for new methods of chemical synthesis. Then he described how he had conquered the seemingly impossible problem of directly combining atmospheric nitrogen with hydrogen through the use of a high-pressure process. He mentioned osmium and uranium as catalysts but did not say a word about the new iron mix.

Haber's speech, one historian wrote, "hit the experts like a bomb." Within days he was approached by a number of companies and individuals who wanted to collaborate, to consult, to profit. Haber received so many requests for reprints of the written version of his talk that he ran out. When he asked BASF to offer advice on how best to provide more, the company replied that they had no interest in helping. "The less that the process . . . is talked about in the nest years, the less attention that is paid by those interested as to whether the technical realization is going to succeed, the more likely it is that we can win an advantage over the competition in the technical use of the process," the company wrote Haber. "And this," he was reminded, "will also be to your advantage."

HABER, FLUSH WITH success, basking in the attention, was ready to move on. He enjoyed getting BASF's money but considered himself an independent scientist, not a corporate lackey. He believed in open communication. Even though he had signed a contract listing the restrictions he would face, it still rankled that he had to ask for permission to publish, had to avoid disclosing anything that might be considered a trade secret, and even had to screen access to his ammonia laboratory. He was now a wealthy, well-known, and much-admired man. BASF was just a step along the way to, he hoped, bigger things. Karlsruhe, his home for a decade and a half, was beginning to feel small and provincial.

It did not take long to find a way out. His wealthy Berlin acquaintance Leopold Koppel, the head of the big electrical firm Auer, told Haber that he was planning something great: the creation of a new institute devoted to pure research in chemistry, to be part of a great German research center being formed in the Berlin suburb of Dahlem. Dahlem was to become an oasis of pure science, an association of prestigious institutes started with money from private donors, employing top talent from academia, and promoting cooperation with industry, all under the aegis of the royal government. Koppel asked Haber to serve as the founding director of what was to be called the Kaiser Wilhelm Institute for Physical Chemistry and Electrochemistry.

Haber did not hesitate. Berlin was the center of German culture, German politics, and German life, and he wanted desperately to be part of it. He accepted Koppel's invitation. In 1911 Haber said good-bye to his friends at Karlsruhe and his business associates at BASF, packed up his house, got on the train with his unhappy wife and his young son, and rode north and east, to the other side of Germany, to Berlin, to fame, and to his fate.

CHAPTER 10

ALWIN MITTASCH HAD a good catalyst, but he kept looking for a better one, one that could spur the ammonia reaction to even greater yields. He tested potential catalysts by the score, then the hundreds, then the thousands. When his efforts ended in 1920, Mittasch figured he had run about twenty thousand experiments. He never found a better catalyst than his iron-aluminum-calcium mix.

The catalyst was important, but it was only the beginning. Carl Bosch was dealing with a mountain of challenges, any one of which could become a disaster. The ammonia process was all tightly connected, one process flowing continuously into the next, which meant that a breakdown at any point could shut down the entire thing. Everything from the input gases to the heaters to the reactors to the recirculation system to the coolers, every connection, every valve, pump, gauge, seal, and fitting, had to work flawlessly, twenty-four hours a day, seven days a week, at high temperatures and pressures. Almost none of the most important equipment existed. It all had to be invented. Systems like this had never been built. Nothing could fail.

The strain was intense, but Bosch seemed able to shrug it off and keep his focus on immediate needs. His next issue was the raw materials for the process, the input gases, nitrogen and hydrogen. Because the new catalyst could be "poisoned" by contaminants, gas purity was vital. Nitrogen seemed relatively easy. A method existed for purifying it from the air in quantity, thanks to Irish drinkers. Fifteen years before Fritz Haber's work on ammonia, the Guinness Brewery in Dublin asked a German inventor for a machine to help keep its brew cool. The answer was an efficient refrigeration machine that worked by taking advantage of a law of nature: gases cool as they expand. The inventor made a refrigeration system so strong it could not only chill ale but also make indoor ice rinks—and cool air into a liquid. Liquid air was a great novelty for scientists, and it made the purification of oxygen and nitrogen possible. Every element has a characteristic boiling point, the temperature where it turns from a liquid to a gas. By carefully warming liquid air the component gases could be separated, boiled off at different points, oxygen at one temperature, nitrogen at another. By starting with liquid air, Bosch could get all the pure nitrogen he needed.

The other component of ammonia, hydrogen, turned out to be a bigger problem. There is very little hydrogen in the atmosphere. The easiest way to get it in bulk was to get it from water (the "H" in H_2O is hydrogen), but the amounts BASF needed were so stupendous that no industrial system existed for making it. Bosch put a team on the problem. It was known that running an electrical current through water could break it apart and release the hydrogen, which bubbled out as a gas. First the BASF researchers tried running electricity through brine, but the process was slow and expensive. They switched to another technique called the water-gas method, which made hydrogen gas by blowing steam over glowing coke. Unfortunately, this method also produced carbon monoxide, a poisonous gas. One of Bosch's team members came up with critical improvements, finding a variation of the water-gas approach that produced massive amounts of hydrogen at a lower cost, while throwing off less carbon monoxide. The hydrogen it produced was not quite pure and Bosch's people figured out how to treat it further, scrubbing it clean of contaminating gases in tall water towers, then running the hydrogen through an activated charcoal purification system.

Gases were difficult to work with. Even with all the improvements, Bosch's water-gas system still generated enough carbon monoxide to pose a danger to workers. Bosch had heard about a copper-containing solution that could clean the poisonous gas out of the system, but his teams found that the solution corroded iron, and just about every piece of machinery they had contained iron. The carbon monoxide issue started turning from a minor problem into a major delay, and it brought Bosch unwanted attention. Heinrich von Brunck had been funneling him plenty of money, but the BASF board was not unanimous in its support of the nitrogen gamble. The carbon monoxide glitch was just another example of how complicated it all was, and what a money-sink it could be. Bosch was reproached by some upper managers who were alarmed at the spiraling costs. Dejected, he turned the problem over to one of the most promising young chemists in his fast-growing team, Carl Krauch, and asked him to find some way to make the copper solution work without corroding iron. After a few months Krauch found a surprisingly simple answer: just add a little ammonia.

They had a catalyst. They had pure input gases. Now Bosch turned his attention to the reactor, the heart of the machine, the high-pressure chamber where hot nitrogen and hydrogen flowed over the catalyst and turned into ammonia. Again, practically everything had to be invented. Haber's tabletop model was no more than a goal, a set of conditions to meet, a hint at structure. As Bosch put it, "There were no examples in industry" to guide his teams.

It was here in the design of the reaction chambers—the workers at BASF called them ovens—where they encountered the real problems. The first was pressure. To build the required pressure Bosch needed compressors bigger and stronger than any ever built. His people reviewed what existed and found that the biggest compressors out there, machines used to pump air deep into mines, were nowhere near powerful enough. Then they looked again at the compressors used for liquefying air. They could be made big enough but they were not built to operate under the conditions needed to make ammonia. The refrigeration compressors used copper as solder for the joints, for instance; the high temperatures near Bosch's ovens would soften the copper and the joints would blow. Refrigeration compressors also tended to be a little leaky. Bosch's machines had to be as close to leakproof as possible because they were working with hydrogen, an explosive gas. So, using the refrigeration compressors as an example, Bosch's teams began to design their own compressors and fittings, employing higher-temperature solders and designing new ways to join metal pieces together so they would hold tightly at enormous pressures.

The reaction chambers had to handle pressures almost twenty times higher than those in the boilers of steam locomotives. They had to run at temperatures high enough to turn iron red hot. They had to be fed carefully controlled amounts of hydrogen and nitrogen gases and kept under constant observation. That meant that ways had to be found to monitor and adjust a variety of factors, from heat and pressure to concentrations of gases and the ammonia product. New sorts of gauges and valves were needed. Ways had to be found to feed in gases and pull out ammonia. How would they heat the chambers? What would they be made of? How could the pressure be maintained with all those inputs and outputs for monitors and gases?

Bosch's teams looked for design hints in locomotive engines, gasoline engines, and the new engine that Rudolf Diesel had invented. Bosch and his engineeers met with men from the German steel industry, learned about the Bessemer process for making steel, talked with Krupps representatives about cannon designs and new advances in metallurgy. He set teams to work designing quick-acting valves, self-closing valves, slide valves; pumps reciprocating and circulating, large and small; temperature monitors of all sorts and sizes; pressure balances; density recorders; trip alarms; colorimeters; high-pressure pipe fittings. Everything had to be rugged, leakproof, functional at high temperature and under enormous pressure. The ovens had the potential of exploding like small bombs; Bosch wanted to make sure they could be carefully monitored and quickly shut down if something started to go wrong. He wanted perfect reliability and lightning speed. He wanted a machine that combined the strength of a sumo wrestler, the speed of a sprinter, and the grace of a ballerina.

He kept hiring engineers to help, doubling the number at BASF in the course of a few years. Bosch was obsessed with the project, working day and night, running his teams hard, keeping the entire picture in mind, seeing systems feeding into systems, fitting his gigantic machine together. Whenever anyone complained about the cost or the pace, whenever anyone balked at working long hours, he would say, "This is about billions!"— billions of marks. The future of the entire company rested on their efforts.

There were enough successes, big and little, to keep the level of enthusiasm high. The catalyst was important. Each new valve and monitor helped. Every time they corrected a design flaw in Mittasch's test machines and prevented an explosion was a victory. Many of the men on the project were dye-industry veterans. They understood that it sometimes took years for breakthroughs. They were not expecting miracles.

Bosch provided them nonetheless. Just a year after Haber's machine first started to work, Bosch decided that he was ready to build a bigger machine. It was a rash move. Nothing, apart from the catalyst, had been perfected. Everything was still under development. Who knew if the oven they had been designing would hold together under pressure? But Bosch refused to wait. He had a good catalyst, he had access to enough pure gases, and he thought enough had been learned from Mittasch's test machines to risk something bigger. Building a bigger prototype was the key, the best way to focus everyone's attention, to get his teams working together effectively, and to uncover problems.

They built the first two ovens from steel cylinders more than eight feet tall, with walls more than an inch thick. They were made by Krupps, Germany's best cannon maker. They were heated from the outside with gas flames, which was worrisome. Earlier tests had already shown that the combination of heat and leaking hydrogen could set off spontaneous spurts of flame. Bosch wrapped the prototypes in reinforced concrete just in case. They were big enough to make hundreds of pounds of ammonia per day—if they worked. His teams called them the big rigs.

Then he lit them up. He looked through a heavy glass viewing port punched through the concrete and watched the bright blue flames as they roared and heated the cylinders. The temperature rose. His assistants turned turned on the compressor and began flowing in the hydrogen and nitrogen. They started measuring, monitoring, and making notes. Then, finally, ammonia started to flow. The big rigs were working, not as well as they needed to work, not up to the levels they had hoped, but they were working.

THREE DAYS LATER they both burst.

Bosch had them taken apart and analyzed. The problem was not in the fittings or the connection points. Everything had held—except for the walls themselves. The inch of steel had cracked. Bosch hid his disappointment. He tried joking about it, saying that it was a good thing they had not used osmium as the catalyst because they would have lost the entire world's supply in one afternoon. He had the fractured cylinders cut into pieces and examined thin sections personally under a microscope. Metallographic analysis was almost unknown in a chemical company like BASF, but Bosch knew metallurgy. He had to know what went wrong and how to fix it. He set up a lab within BASF for the testing and improvement of metals.

What he saw under the microscope required some thought. Steel from the outside of the cylinders, the side farthest away from the ammonia, looked normal enough. But the inside metal, the steel close to the high pressure and heat, was swollen and shot through with hairline cracks. Strength tests showed that the inner metal had lost elasticity. It was turning brittle, like a piece of bread dried in an oven. Bosch ordered more studies. The brittlization, as he called it, started on the inside surface and, as operation continued, slowly ate through the wall until the remaining steel was so thin that the pressure blew it out. This had not happened before. Mittasch's score of little ovens for catalyst testing, for instance, had run for months without the reaction chambers becoming brittle.

Perhaps it was the heat. They were blasting the containers with flame, heating from the outside, and perhaps the resulting temperature gradients strained the metal. They played with the heating system, moving the heat source inside the ovens, trying different ways of cooling the metal, increasing the insulation, juggling the ways they used the hot exhaust gases to preheat the cold incoming gases. The heating system became more efficient. But the steel kept cracking.

Perhaps ammonia was the problem. As Bosch well knew, ammonia could react with hot iron to form iron nitride, the same substance that proved the undoing of Wilhelm Ostwald's machine. Iron nitride was brittle. Perhaps his ovens were being contaminated with it. Somewhat to Bosch's surprise, however, chemical tests showed no significant amounts of nitride in the damaged metal.

His teams kept testing. The oven walls were made of Krupps's best carbon steel, the strongest available. Carbon steel is an alloy of iron and carbon, but Bosch's tests uncovered the fact that in the most damaged areas of the oven walls, the carbon was disappearing from the steel. Once the carbon was gone, only soft iron would be left—which is much weaker than steel—but the walls were not softening; they were growing brittle. They were not turning into pure iron. They were turning into something else. Finally, they found that the brittle metal was shot through not with nitride but with something they had not expected: hydrogen. It looked as if the hydrogen gas inside the oven was somehow being forced into the metal, reacting with it in some way, weakening it, making it crack.

"We were then in a dilemma," Bosch remembered later, with characteristic understatement. No one had seen a chemical reaction like this before, much less figured out how to prevent it. The only thing they knew for certain was that hydrogen was connected with the problem. The detailed chemistry would take years to untangle, but Bosch did not have years. Unless he could come up with a practical solution, and quickly, the game was over.

It was undoubtedly related to hydrogen's behavior under high temperatures and pressures. But he could not lower the temperature (it was needed to break apart the N_2) or change the pressure (it was needed to push the formation of ammonia) or get rid of the hydrogen. Hydrogen was everywhere in the oven. There was no way to eliminate it.

That left only two options: find a new metal that would resist the hydrogen effect, or find a way to protect the existing steel. It was hard for Bosch to imagine what a new wall might be made of. The issues were strength and cost. Platinum and other rare metals were too expensive. At the extreme pressures and large oven size they were using, carbon steel was the only practicable thing in 1910 strong enough to hold together. Bosch's metals experts began looking for ways to improve it, to create a more hydrogen-resistant carbon steel, flavoring it with other elements, adding dashes of molybdenum, tungsten, chromium, anything that might offer an improvement. Some of these new alloys lasted longer, but eventually hydrogen attacked them all.

Their last hope was to protect the wall, to line it with something high-pressure hydrogen could not get through. They started testing liners, but every material they tried either let hydrogen through or was itself attacked. The high pressures seemed capable of forcing the small hydrogen atoms into and through everything. They considered gilding the insides of their ovens, applying liners of pure gold, but did not test it. Even if it worked it would be too expensive.

Six months after the big rigs burst, Bosch still had not found a solution. He now had scores of scientists and engineers working on the project, supported by hundreds of assistants and laborers. None of them could find an answer. Every oven they tested exploded within a few days.

Bosch took his usual Friday after-work break, a night of bowling and beer, his one luxury. He enjoyed socializing with his team at the end of the week, forgetting about work or at least talking about it in a lighter way, getting to know his team leaders. It helped build team spirit. And he really liked beer.

The next morning, he realized they had been going about it all wrong. They had been trying to change the steel, to protect the steel. Why not simply accept the fact that hydrogen was going to attack it? There apparently was nothing they could do about it. High-pressure, high-temperature hydrogen was like some sort of universal solvent for metals; it eventually got into them all. Once he accepted that fact, Bosch realized something else. They had been asking two things of their oven walls: first, that they contain enormous pressures and, second, that they prevent the escape of gases, particularly the explosive hydrogen.

What if he separated those two things?

Bosch quickly jotted down some ideas. He was thinking about how the high-pressure hydrogen was pushed into the metal. What if they designed a new kind of liner, one made not to stop the hydrogen but simply to absorb the damage, a sacrifice thrown in front of the hydrogen, a weak shield made to take the hit and save the strong steel wall?

Let the hydrogen slowly diffuse through the liner, change it, and weaken it. Whatever gas reached the other side would arrive at a much lower pressure and would be much less likely to attack the strong steel outer wall. The liner would not have to be some elegant, strong, expensive material. A common metal would do, something like soft steel. Fit it tightly inside the strong steel outer cylinder and let it grow brittle—if the fit between liner and outer case was tight enough, there would be no place for it to go, no room to expand, and no chance of big cracks. It might last for months. When the liner's life was up, shut the oven down for a few hours, strip it out, and put in a new one. It would be fast and cheap.

If it worked. Bosch's teams started testing. Bosch had the inside surface of the thick, carbon-steel wall grooved so that any gases trapped between the liner and the wall could circulate and be removed. He was still bothered by the thought that explosive hydrogen might collect and detonate. When he was walking to work one morning, another answer came to him. Why not let the hydrogen out? They had all been so focused on making the outer wall impermeable, holding in the gases and the pressure. But what mattered was not whether hydrogen leaked but how much. If only a little leaked, and leaked slowly into a large area, the concentration would not build up enough to pose a danger. Early tests of his inner liner idea looked positive. The inner liner would help contain the gases and the pressure, which meant . . .

He sat on some nearby scaffolding and made a rough sketch of the thick outer wall pierced with rows of holes, small enough to maintain the general integrity of the outer wall, big enough to let out any remaining hydrogen between liner and outer wall. He took his drawing to the patent office. It was another low-tech, low-cost, Bosch-style solution. His workers drilled scores of little perforations, each about as big around as a pencil, through the once-impregnable carbon-steel outer wall. They called them Bosch Holes.

The liner and the holes were smart solutions, cheap and easy, and they worked. The pressure was contained. Hydrogen loss was minimal. Bosch's teams started building and testing full-sized ovens, and they stayed intact after weeks, then months. The project was back on track.

There were still occasional problems, flares of gas, and breakdowns at some point in the interlocking systems. Occasionally an oven exploded, but this was now rare. Bosch now had created at BASF a complete materials testing laboratory, lavishly equipped with the newest and most sophisticated equipment, surpassing even the best labs in the steel industry. Whenever anything went wrong with metal, the problem could be analyzed quickly and efficiently. His people found that the strength of most steels started to drop when temperatures reached 300°C (about half the temperature inside an ammonia oven) and fell alarmingly, by up to half, at temperatures above 500°C. That meant that the temperature gradient across the oven walls, from the inside with its 600°C temperatures to the cooler outside, could stress steel unbearably. Researchers were set to work to find ways to spread the heat as evenly as possible, to lessen the stress through improved ways of applying flames or electrical heat to the reaction, through faster ways to get hot reactants out, or better ways of cooling. Other teams worked on the heat exchangers, which included copper tubes that were also destroyed by hydrogen, and found a way to use N_2 to flush hydrogen out of that area between the sleeve and the outer wall. Other teams kept working on ways to increase energy efficiency and decrease gas loss, on new meters and valves to control and track the increasingly complex flow of gases through the new design, on gaskets, and flanges. Every piece was constantly reviewed, every system constantly improved.

Gradually the breakdowns became fewer and the ammonia yields crept higher. Bosch's teams were getting to know their machine. By the end of 1911 the prototype ovens were running flawlessly for long periods of time, producing tons of ammonia at a cost cheaper than anything else on the market. The research had been expensive, but as problem after problem was solved, the final system was highly efficient. Bosch's solutions were not high cost, idealized, or theoretically impeccable; his machine was not a perfect machine. It was a series of practical fixes, cleverly and solidly stitched together. And it worked.

By early 1911—much faster than most observers had thought possible—Bosch's BASF prototype plant was turning out more than two tons of ammonia a day. Plans were made for additional processes to turn the ammonia into fertilizer ready for farmer's fields. Plans were made for taking the final step: even bigger ovens, a full-scale factory, and full-scale profits. BASF began looking for a site, preferably near water to help with hydrogen production. Company men rode through the nearby countryside, walking through large acreages of farmland on the Rhine. They found a good site just a few kilometers north of the firm's main plant at Ludwigshafen, at a village called Oppau, and started drawing blueprints for the world's first synthetic nitrogen factory.

Bosch had now expanded his research teams to the point where, as he put it, "It is probably true to assert that such numbers have never before been engaged on one single problem." Under Bosch's direction BASF's nitrogen project was growing into the biggest scientific effort in history, comparable in scale to the Manhattan Project in World War II.

It was then, in September 1911, that BASF received word

that its basic patents regarding nitrogen fixation were worthless. A lawsuit to nullify them was being filed by a huge competitor chemical firm, Hoechst. The suit asserted that Haber's "discovery" was not that at all; Nernst had already shown, more than a year before Haber, that nitrogen and hydrogen could be combined under pressure to make ammonia. Ostwald himself was advising Hoechst in this regard. The facts were in the scientific record; Walther Nernst and Fritz Haber had argued about them in public at a meeting of the Bunsen Society some years earlier. BASF continued to move ahead on the plans for the Oppau plant while putting a team of lawyers to work on the nullity suit.

Two months later, on December 4, 1911, Bosch's friend, mentor, and boss, the most important backer of BASF's nitrogen gamble, Heinrich von Brunck, died.

Suddenly, it seemed, all bets were off.

CHAPTER II

S UING WAS A clever move. Hoechst, another of Germany's
biggest dye and chemical firms, was trying to strangle
BASF's baby while it was in the cradle by either nullify-
ing the Haber patents, robbing a competitor of exclusive rights
to a moneymaking process, or at the very least forcing BASF to
divulge valuable details of its nitrogen research in order to fight
the suit. Whatever happened, Hoechst came out ahead.

BASF had kept Carl Bosch's nitrogen work as quiet as
possible—nothing public had gone out since Fritz Haber's talk
in 1910—but Hoechst had been keeping tabs. As far as Hoechst
was concerned, watching quietly on the sidelines as BASF
staked sole claim on an entirely new and hugely lucrative field
was unacceptable.

Haber was notified about the suit in early September 1911,
just after assuming his position as head of the new Kaiser
Wilhelm Institute for Physical Chemistry and Electrochemistry. To allow himself to be lured to Berlin (just where he
wanted to be) he had negotiated a munificent salary, a villa for
himself and his family, a chair at the University of Berlin, mem-
bership in the Prussian Academy of Sciences, thirty-five thousand marks a year for the physical upkeep of his institute, and
three hundred thousand marks a year for its operation. He was
eager to create his corner of the scientific Eden in Dahlem, a

place devoted to free inquiry and great developments in the name of the kaiser. The Kaiser Wilhelm Institutes (KWI) were something relatively new in science: a government-sponsored (although mostly privately funded) site for research, intended to draw the best minds, to provide them a home and support free from the duties of academia and the restrictions of industry. It was modeled in part on the Carnegie Institute in the United States. It was a noble and patriotic project, part German imperial showcase, part scientific foundation, underscoring the pre-eminence of German science and magnifying the glory of the German state. For Jews like Haber, it was also an alternative to high military service (from which they were barred) as way of showing devotion to the state. Jews, who constituted only 1 percent of the German population, gave 35 percent of the funding to start the KWI. Jews, baptized and unbaptized, were employed in higher proportions at the institutes, particularly Haber's, than they were at most universities. Here, in science, in Dahlem, Jews would be given the chance to contribute to German culture and German progress. Here was a chance to show that race did not matter. Here, it seemed, was the future. Many scientists who worked in Dahlem considered the KWI "the emperor's academic guard regiment." Holding a directorship, as Haber did, was a signature honor.

Haber, his wife Clara, and their son Hermann arrived in Berlin in the middle of the summer. There, one of his best friends later wrote, "Fritz Haber completed his transformation from a great researcher to a great German." He was, his son wrote, "the embodiment of the romantic, quasi-heroic aspect of German chemistry in which native pride commingled with the advancement of pure science." His nitrogen success had raised his reputation enormously. There was already talk of a Nobel Prize. At the institute, he used his renown to establish total control, insisting on a free hand in all hires and projects, total decision-making power over patent applications, the continued ability to offer paid advice to industries that might seek it, and—in recognition of his continuing health problems—permission to disappear as often as needed for visits to visit spas and health retreats.

Haber thrived at Dahlem. He was finally at the center of things, an unabashed German patriot to an extent "unusual," his son noted, "even in an age when jingoism . . . was condoned." He was near the heart of German government, his opinion asked, and his expertise unquestioned. Thinking that he might be invited to the palace at any time, he studied court etiquette in his office. Once, a colleague remembered, Haber was walking backward in his office, practicing bowing his way out of an imperial audience, when he knocked over and broke an expensive vase.

The nullity suit hit just as he was settling into his new life. He immediately asked BASF what to do, and the company responded by asking Haber to immediately transfer all remaining nitrogen-related patents. He did. Then BASF started making overtures to Haber's old nemesis Walther Nernst, the man whose work was cited in the Hoechst suit. Nernst was invited to visit the BASF laboratories to study Bosch's progress and to judge for himself if what they had accomplished was patentable. The company provided Nernst with access to a laboratory, equipment, and chemicals.

The case was set to be argued before a national court in Leipzig in early March 1912. Two days before it commenced, Bosch met with Haber, Nernst, and BASF research director August Bernthsen. They talked about where they were with the case. They were not sanguine. Records of the meeting show that, as one historian put it, "they judged their position as weak and unfavorable."

Hoechst's representatives, on the other hand, were optimistic. As the court proceedings started they read into the record an expert opinion from Wilhelm Ostwald in which the great chemist dismissed Haber's results as a pure extrapolation of known concepts, a mere application of increased pressure to a system that had been known about and explored since his own work around 1900. Earlier research in this area, especially Nernst's, rendered Haber's so-called discovery "not only scientifically probable but . . . scientifically certain." This was the heart of the Hoechst case: Everything Haber did was based on something Nernst had already done. Haber's patents, therefore, should be rendered null. They took their time mounting a thorough justification of their company's suit. When it came time for BASF to respond, the Hoechst side was delighted to note, Bernthsen seemed able to offer nothing more than a brief answer.

Then the door opened and Nernst himself appeared arm in arm with Haber. The two men gave every impression of being best friends. Then Nernst, the man at the heart of the Hoechst case, made what an observer remembered as a "passionate speech" to the court. In his opinion, Nernst said, his own earlier work in this area had no significant technical relevance to the advances made by Haber. Haber, he said, had broken new ground in his exploration of new, higher-pressure ranges. Nernst's recent experiments had convinced him that the original Haber patent dealt "with results of a completely new type, and the declarations therein form a solid experimental foundation for an extremely important new technical process. For this reason it seems to me without a doubt that the invention described in the patent under discussion is to the fullest measure worthy of the protection given by the granting of a patent."

As he spoke, a Hoechst representative leaned over to his lawyer and whispered, "We might as well go home."

A historian later noted that "Nernst's change from sharp critic to well-wishing supporter of Haber was amazing." Or perhaps not. In the weeks prior to the trial, Nernst signed a five-year contract with BASF, which had agreed to pay him an honorarium of ten thousand marks a year to serve as an expert consultant.

Later that afternoon, Bosch and the BASF team wired corporate headquarters: "Hoechst's nullity suit against our NH_3-pressure patent rejected and they are to pay costs."

WHATEVER WORRIES HABER and BASF might have had now evaporated. Clearing the Hoechst suit also helped quell the worries of those members of the BASF board who, now that Heinrich von Brunck was dead, might have returned to their arguments against the nitrogen project. Even more important were Bosch's recent advances in production. Ammonia was reliably flowing by the ton from his prototype machines. The system worked, and even without Brunck's urging, the board had little choice but to take the final step and approve the construction of a full-scale ammonia factory. They had invested too much money to do anything else.

Bosch was put in charge of the construction and on May 7, 1911, the first spadeful of dirt was turned at Oppau. Bosch oversaw every aspect of design and construction, instituting a system of total planning that knitted together, from beginning to end, all the disparate parts of the complex project. Oppau would have absolutely everything: ranks of compressors the size of locomotives, ovens four times bigger than the big-rig prototypes, a mini-factory for freezing nitrogen out of the air and boiling it clean, another for turning ammonia into fertilizer, miles of pipes and tubing, a complete electrical system with its own generators, a complete shipping center with its own rail yard, a research laboratory with 180 chemists and a thousand assistants, and housing and transportation for more than ten thousand workers. Each element would feed into the next, a continuous flow of people, power, and raw materials. Bosch was creating a single machine as big as a town.

He did not care much about how things looked, just how they worked, yet his buildings ended up looking remarkably modern, more like industrial designs from the 1950s than those of 1911, with simple, clean lines, large open areas, and large windows for natural light and ventilation.

Oppau opened just sixteen months after construction began, in September 1913. There was a shakedown period with breakdowns and delays but nothing Bosch's well-honed teams could not handle. The bigger ovens, fitted with all the improvements they had discovered in the previous three years, worked beautifully. Within a year Oppau was making tons of ammonia-based fertilizer every hour, pricing it competitively with the Chilean product, and selling it as fast as they could make it. Once it got into full swing, the profits were enormous.

As SOON AS it became clear that Brunck's and Bosch's ammonia gamble was going to pay off, BASF dropped all of its other nitrogen-fixing projects, halting the development of Otto Schönherr's arc furnace and abandoning its deal with Norsk Hydro. The high-pressure process was better than anything else out there. The Oppau plant used far less energy than the arc furnaces in Norway, did not depend on nearby waterfalls for electricity, and could be scaled up indefinitely. Bosch had done everything he could to maximize efficiency and cut the cost of production. The ovens at Oppau recaptured so much of the heat released by the formation of ammonia, for instance, that they hardly needed any added power; once fired up, they were almost self-sustaining. Building the plant was expensive, but once the capital costs were paid, it took very little money to make ammonia. The main cost was still producing pure hydrogen. But Bosch was working on that.

Oppau had barely reached full production before Bosch was planning an expansion. Instead of seeing Oppau as a final stop in the development process, Bosch seemed to view it as another, bigger prototype, a step on the way to even better things. The ovens at Oppau were big, but Bosch told his engineers he wanted them bigger, taller, and wider. "We need to go to the limits of the German steel industry," he said.

That limit, in 1913, was a cylinder—it looked like a big iron pipe—about twenty feet long. Even German steelmakers could not cast a strong enough chamber any bigger. So Bosch decided to attach two of them together end to end, finding a way to angle the bolts so that the joint would hold under the pressure, doubling the size of the ovens (and the amount they could produce) without asking more from the foundries.

With the ovens at their maximum size, the catalyst as good as it could be, and all the ancillary processes, the gas production and fertilizer making, working as well as they could, the only additional need was labor. Bosch boosted the workforce at Oppau 20 percent in two years. He ran his factory twenty-four hours a day, with laborers working overlapping nine-hour shifts. The oven stopped for only quality checks, repairs, and liner replacement. He built more housing and arranged train schedules to ease the arrival of thousands of laborers who had to live at a distance. The stream of ammonia grew into a torrent.

The speed at which Bosch made things happen was surprising, even to the BASF directors. High-pressure chemistry had never been done at the industrial level. It was terra incognita for Bosch and his teams, and problems had to be solved in new ways, with new equipment and new ideas. They invented as they went. As a result, Oppau was not just an ammonia factory. It was the birth of an entirely new technology, high-pressure chemistry. BASF patented it all, generating a blizzard of applications for the ovens, the catalysts, quick-acting magnetic valves, continuously functioning gas flow meters, devices for measuring high-pressure gas densities, novel heat exchangers, recycling systems, new gaskets, the list went on and on. No one else in the world could do this work.

Bosch began thinking: If this high-pressure technology could make ammonia, what else could it make?

• • •

IN THE SUMMER of 1913, Nernst and the gray eminence of German physics, Max Planck, took the night train to Zurich. Their goal was to lure back to Berlin the greatest physics theoretician of their age, the oddball genius Albert Einstein. Einstein had been born in Germany but left as a young man to protest the country's militaristic culture (and to avoid his own compulsory military service). They hoped to entice him back to Germany to become the centerpiece of the KWI, a crown jewel among his native nation's researchers, another proof of the greatness of German science.

In Zurich the two men made Einstein an offer he could not refuse. If he returned to his native land he would not only be given a substantially increased salary and research support but also would be made the youngest-ever member of the Prussian Academy of Science. "You know," Einstein told a friend, "the two appeared to me as if they were people who wanted to get hold of a rare postage stamp." He had little respect for Germany, but the new post would free him from the teaching responsibilities he carried in Zurich, give him more time to ruminate—and put him nearer to his German cousin Elsa. Einstein's marriage was falling apart. He had a crush on his cousin.

Once he arrived in Berlin, Einstein came to know Fritz Haber. Haber had pushed for bringing Einstein back to Germany; he greatly respected Einstein's mind. But the two men could hardly have been more different. Haber was completing his transformation into the perfect German: promilitary, pro-kaiser, ever more the stiff-necked Prussian-style patriot. Einstein was a freethinking, wise-cracking cosmopolitan, a bohemian pacifist who enjoyed making fun of precisely the traits Haber was coming to embody.

Einstein was perfectly capable of puncturing Haber's growing egotism. "Haber's picture unfortunately is to be seen everywhere," Einstein wrote in 1913 while settling into the KWI. "It pains me every time I think of it. Unfortunately, I have to accept that this otherwise so splendid man has succumbed to personal vanity and not even of the most tasteful kind." Nevertheless the two men respected each other as scientists. They also shared the experience of being among their nation's most prominent Jews. Regardless of their achievements, they would both be perennial outsiders, eternal second-class citizens. Einstein understood that, and hated it. Haber still labored for acceptance.

Yet the two men struck up a strong friendship. Einstein tutored Haber's son in mathematics, and Haber helped Einstein through the disintegration of his marriage. When Einstein's first wife finally got tired of his infatuation with his cousin and left him, Haber accompanied the distraught Einstein to the train station. After Einstein said good-bye to his wife and two sons, Haber kept the weeping young scientist company through the night as he wept. "Without [Haber]," Einstein said, "I wouldn't have been able to do it."

Einstein might have been unhappy as well about the state of the world. The day before he said good-bye to his family, Austria had declared war on Serbia. A few days later, Germany declared war on Russia. Einstein, during that long night after his wife left, talked with Haber about the military mind-set of Germany and the madness of war. Haber, on the other hand, had already volunteered for service.

CHAPTER 12

C ARL BOSCH SPENT the first half of 1914 coaxing the Oppau plant into full production, smoothing the last glitches, fine-tuning the system, and optimizing output. By summer, in the last warm weeks before the war, the factory was producing as efficiently as he could make it. The fertilizer market was his primary concern, and to master it he started yet another research team, this one devoted to understanding how fertilizers worked and how to make them work better. He built a research farm and laboratory in a field near the Oppau plant to test his products and improve them. The easiest fertilizer to make from ammonia was, in chemical terms, ammonium sulfate. Many farmers still preferred the Chilean sodium nitrate they had used for years, and Bosch wanted to prove to them that his product was just as good, if not better.

The appearance of the Oppau plant established the idea of synthetic fertilizer in people's minds, produced through what people were starting to call the Haber-Bosch system, crediting both the initial researcher and the man who turned it into an industrial reality. The hyphenated name also symbolized a shift in scientific reality. In the twentieth century, industrial laboratories and industrial applications would increase in importance, until industrial researchers ranked in importance with academic scientists.

The Haber-Bosch system still had a long way to go. In 1914 it was limited to a single, highly efficient plant, far short of what would be needed to supply Germany's needs, much less the world's. In terms of pounds of fixed nitrogen, it lagged far behind imports of Chilean nitrates (as long as they continued) and Norwegian arc-process fertilizer. A group of influential German businessmen and bankers was heavily backing a competing technology, one that used huge amounts of electricity to make calcium cyanamide—another nitrogen-containing fertilizer—from high-temperature reactions between N_2 and calcium carbide. It was called the cyanamid process. In the United States the technology was used by American Cyanamid, a giant chemical company that tapped electricity generated at Niagara Falls. In Germany a consortium of investors was trying to get the government to invest in more cyanamid plants. To compete, BASF needed to build more Haber-Bosch factories.

There was plenty of demand for everybody, but BASF thought, rightly, that it had the best technology. Oppau was already making big money—BASF had started making a profit from nitrogen in 1913—and the demand for fertilizer was so high that the plant could sell everything it could put out and more. Bosch wanted to immediately expand Oppau. But construction costs were so high that the BASF board hesitated.

Then the war started. The Germans, expecting a brief and glorious march to Paris, figured that their stores of explosives and raw material for making more—primarily in the form of stockpiled Chilean nitrate—would last about six months. That, they thought, was plenty of time to win the war with powder to spare.

Fritz Haber, perhaps also expecting a short war, made certain that he entered military service before it was over. He was too old to carry a gun, and too Jewish to be an officially commissioned officer, but he could be of service in other ways. The war offered him another chance to serve his nation, to prove how indispensable he was to the government, to move closer to the palace, to advise the kaiser. He took a role as scientific adviser to the government, with a focus on ways research could help the military. And he quickly realized that if Germany was wrong about the length of the war, if it went on for a few extra months, his nation would face a critical shortfall of the basic raw material for explosives: nitric acid, which arms makers used to create both gunpowder and high explosives like TNT and nitroglycerine. Nitric acid was the be-all and end-all of arms production. It could be made relatively easily from both Chilean nitrate and the cyanamid process—a point that Germany's cyanamid industry began stressing with the government—but not from ammonia. In late August Haber wrote BASF asking if the firm was capable of some sort of large-scale conversion that would turn his ammonia—"his" in the sense that he still made a few pennies on every kilo—into nitric acid. The company responded that the idea was "an impossibility." But Bosch and others began thinking about it. Nitric acid was about to become extremely valuable. Haber's letter triggered high-level discussions at BASF. Bosch and his board members had been thinking about fixed nitrogen for food; now they started thinking about it for bombs.

The need quickly became critical. The Germans invaded France through its relatively undefended border with Belgium, a move that caused an international outcry because Belgium was officially a neutral nation. The image of German soldiers abusing neutral Belgian citizens became standard propaganda for the rest of the war. Somewhat to the Germans' surprise, the French, joined by the British, fought back furiously. At the Marne River near Paris, a few weeks after the start of the war, the Allies dug in and stopped the Germans. What was to have been a lightning war bogged down into a slugfest in the trenches, with progress of the German invasion measured in feet, under cold rains as summer turned into fall. Both sides began burning tons of explosives as they turned to artillery barrages to destroy fortifications and spray dug-in troops with shrapnel. The German military began scouring the nation for Chilean nitrates, confiscating industrial holdings, raiding warehouses, and looting agricultural depots. They bought all the nitrates they could from neutral countries. They got a break when they captured Antwerp and found a warehouse full of thousands of tons of Chilean fertilizer. It was enough to keep them going but not for long. In November 1914 the Germans estimated that twenty-nine kilotons of fixed nitrogen would make enough gunpowder to supply the military through an entire year of warfare. A year later, as the war dragged on, they were burning that much every ten weeks. A year after that, with the war still going, every five weeks. The side that won would be the side with greatest access to fixed nitrogen.

CHEMISTS BECAME VERY important. The government started dealing with the cyanamid industry, funding an expansion of their plants as a way to get more explosives. Haber thought that was a mistake for at least two reasons: first, because the process was inefficient compared with Haber-Bosch, requiring both more laborers per ton of fixed nitrogen and consuming more power, and second, because he personally would make no profit from it. The only way for him to benefit from the wartime need for fixed nitrogen was to find a factory-sized way to turn Haber-Bosch ammonia into nitric acid. Haber talked with BASF about it. A method was available—it had been patented some years earlier by the omnipresent Wilhelm Ostwald, using platinum as a catalyst—but there had never been much of a market for it. As far as they could see, only a single small ammonia-to-acid factory had ever been put into production in Germany. Profitability had been hampered by the cost of platinum, a catalyst so expensive that it was hard for Ostwald's method to make much money. Now, with the growing need for nitric acid, Bosch decided to find a better way. Mittasch ordered his well-equipped team to look for a replacement for platinum, something just as good but cheaper. They quickly found one, another iron mix, this time using bismuth as a promoter. BASF had a catalyst, but that was all they had when at the end of September, shortly after the war bogged down, Bosch met with the War Ministry. BASF needed to make a move. The government was desperate, and the cyanamid lobby had been busy. They kept repeating how easy it was to make explosives using the cyanamid process, and it was true—unless a better way could be found to convert ammonia, cyanamid was probably the best choice. There was a great deal of interest in spending government money to build more cyanamid plants, and plans were made to start them while Bosch looked for a cheap way to make nitric acid from ammonia.

He could not do it. Time was running out. But then he thought of another way. It did not have to be a direct conversion to nitric acid. What if the ammonia went through a middle step, a step with other benefits? He knew of a way to transform Oppau's ammonia into something almost as good a nitric acid: Chilean saltpeter—the Germans also called it white salt—the same sodium nitrate the nation imported by the shipload from South America. The methods were already proven, the machinery buildable. It would be expensive to build the white salt conversion plants, but this was wartime, and the effort would undoubtedly be supported by the government. It would allow BASF to tap into the kaiser's war spending. It would keep the company in the race with the cyanamid plants. Perhaps it could even pay for the expansion of Oppau.

At the meeting with the War Ministry, Bosch made a rash promise: He pledged that within six months of the day of the meeting, the BASF ammonia plant at Oppau could be refitted to produce five thousand tons per month of white salt, which in turn could be easily transformed—exactly as Chile saltpeter already was—into both gunpowder and, through another process, nitric acid. Then he laid out the reasons why his plan would be cheaper than the cyanamid industry's.

It was a lot to promise, but BASF needed to think big. Government funding offered a way to help pay the costs of expansion, and expansion is what it needed to compete. The chemical company also knew that the primary form of fertilizer it had been producing, ammonium sulfate, was still less attractive to farmers than Chilean saltpeter. The capability of converting its ammonia not only opened up the arms market but would help it offer a new product when the war was over. Once it could make the Chilean product in its own plants, Germany would be at last completely free from the South American trade and its risks—as Ostwald had hoped—and BASF would reap the profits.

With Haber's smoothing the way within the military, Bosch made a deal with the German government that became known as the Saltpeter Promise. BASF would start making Chilean saltpeter, starting with five thousand tons per month, as Bosch had promised, and quickly increasing to seventy-five hundred tons per month. In return, the government would give BASF six million marks in support to expand and extend production at Oppau and to build the needed conversion plant alongside the ammonia works.

When the deal was signed, BASF was no longer just a chemical firm. It was a defense industry. Bosch did not much like it. His team recognized the irony: They had worked long and hard to feed people; now the same technology was going to be used to kill them. Bosch did not talk about it much, but he felt it. One of Bosch's top assistants remembered that during the saltpeter negotiations Bosch had referred to "this dirty business." When the deal was done, he said, he was going to drink himself into "the biggest high of my life."

. . .

UNTIL THE NEW conversion plants were up and running, however, Germany still had to rely on shipments of nitrate from South America. This led to one of the stranger moments of World War I when, on November 1, 1914, the first major sea battle of the war began—halfway around the world from Germany and France, off the coast of Chile. It was brief and brutal. In heavy seas, with darkness falling, a squadron of Germany's most modern warships led by Admiral Maximilian Graf von Spee engaged and sank several older, badly outgunned British warships. The battle continued by moonlight, with the Germans aiming at the fires that were burning on the British ships. The British lost two cruisers and sixteen hundred sailors and officers. The Germans did not lose a ship; their casualties totaled two wounded. It was the Royal Navy's first significant defeat since the days of Napoléon.

More important than the blow to British pride was the practical result: The Germans, during the critical early months of the war, cleared the British navy from west coast of South America. Germany, at least for the moment, controlled the shipment of nitrates from Chile. Spee's success was so total, the German danger to British shipping so great, that insurers refused to extend coverage to British nitrate ships. The British depended on the Chile trade for their gunpowder and explosives too, and Spee's imposition of what amounted to a German blockade started exerting a slow stranglehold on the United Kingdom's war-making capability. As a U.S. military expert of the day said, "To strike at the source of the Allied nitrate supply was to paralyze the armies in France. The destruction of a nitrate carrier was a greater blow to the Allies than the loss of a battleship."

It provided Germany a respite while the government raced to get its own nitrate plants started. But it did not last long.

Within weeks the Allies dispatched a powerful squadron to hunt down Spee. Knowing that superior forces were en route and that any help from home would arrive too late, the admiral tried to make a dash back to Germany while he still could, leading his ships around Cape Horn, heading for the north Atlantic. On the way he needed fuel, which led to a raid the British coal bunkers in the Falklands. It was a move the British had foreseen. On December 8, 1914, the British opened fire and blew the Germans from the sea. Among the nearly two thousand dead were Spee and two of his sons.

For the rest of the war, the Allies maintained firm control of the Chilean nitrate trade. The number of Allied freight vessels carrying Chilean nitrate quickly doubled. Germany, on the other hand, was now completely cut off from both its major source of arms and its major fertilizer for food. The need for home production became that much more important.

As Spee was racing for home, German field commanders were warning Berlin that a shortage of gunpowder threatened their success on the western front. Whatever was going to happen with powder and explosives needed to happen fast.

With his infusion of government money in hand, Bosch assembled his teams, brainstormed with his engineers, and on October 24, 1914, started building a system to turn ammonia into Chilean saltpeter. The money allowed them to design and build a new generation of ammonia ovens, twice as big as the original models, giants nearly forty feet tall lined with soft steel, dotted with Bosch Holes, and packed with sophisticated, ever-improved high-pressure technology. The bigger the ovens got, the more efficient they became. Bosch's new ovens confirmed his technology's position as the world's most cost-effective way to fix nitrogen.

In May 1915, just eight months after he made the Saltpeter Promise, Bosch started making good on it. White salt began to flow out of Oppau at the rate of 150 tons a day. The company capitalized on its success, bringing in politicians and military officers for tours of the great plant at Oppau, with its ranks of giant ovens, its enormous compressors, cadres of scientists, thousands of laborers, all working night and day for German victory. Press releases were sent out, conferences were held. Oppau became a point of pride and a tour destination for the wealthy and powerful of Germany, high government officials and top industrialists. The facility's size, power, and efficiency impressed everyone. The idea of making war out of air and water—something the Allies had no way of doing—was impressive. Bosch's giant Oppau factory, operating twenty-four hours a day, became its own best promotion tool. Using Oppau as a model, BASF outflanked the cyanamid interests, building the impression within government and in the press that the Haber-Bosch process was the only nitrogen-fixing system worth backing.

They were right: It *was* the best available way to make explosives. It soon became clear that the cyanamid process required more electricity and more workers per ton of fixed nitrogen than Haber-Bosch did. The cyanamid plants were less efficient and more expensive. After 1915 the German government stopped any further investments in the cyanamid approach, and pursued only Haber-Bosch. As Oppau grew and production increased, the method took an increasingly large share of the market. But Oppau was just the start.

No one at Oppau was expecting an attack from the air. The whole idea of airplane warfare was too new, the techniques too primitive, to imagine that the enemy would try to fly into Germany to attack a factory. But on the sunny morning of May 27, 1915, that is exactly what happened. A squadron of French planes, small, flimsy machines made of canvas and wood, buzzed over Oppau and began to drop small bombs on the buildings, the trains that carried workers to the plant, and the workers themselves. Oppau had no air defenses, but damage was slight (mostly because bombing technology was so rudimentary). The event did, however, mark the beginning of a new kind of war, an air war directed not at combatants but at the factories that supplied them.

A machine-gun post was set up outside the plant. Searchlights and new antiaircraft guns were added. BASF built a sham chemical plant near Oppau to lure the flyers away (no one was fooled). The raids continued, including, if the moon was bright, nighttime bombing runs. Bosch's genius in making his factory one great integrated machine was good for efficiency but bad during wartime. Oppau was so closely knit that one well-placed bomb could shut down the whole thing. Minor damage that shut down the factory for a short time could do major damage from repeatedly turning off and restarting the ovens, heat exchangers, and gas compressors. Over the course of the war, Oppau suffered more from its own repairs (done with inferior wartime steel) and restarts than it did from enemy bombs. As the conflict went on and the planes continued to buzz, the plant began wearing out. Something had to be done.

There was no good way to stop the attacks. It was a matter of geography: Oppau (like the rest of the BASF plant at Ludwigshafen) was close to France, on the west side of the Rhine, within easy reach of the French planes. The Germans could not afford to lose the plant or even suffer a significant decrease in its production. In September 1915 the kaiser's government proposed another deal to BASF. For the good of Germany the company would build a second Haber-Bosch plant more than twice the size of Oppau, far from the Rhine. Bosch suggested a site in the center of Germany, close to good stocks of coal and water. Negotiations went back and forth for months, with Haber acting as intermediary. BASF was concerned that a plant this size would create a glut in the ammonia market when the war was over. It wanted the government to shoulder the entire cost. The German government responded with a proposal to make the plant a German national facility, with BASF hired to operate it. But with stocks of ammunition eroding day by day, the government was in a poor position to bargain. It was finally agreed that BASF would own the plant, which would be built with a government loan of thirty million marks to be repaid after production started, plus other attractive supports. The final deal was struck while the long, bloody Battle of Verdun was under way.

A final site was chosen. The plant, it was agreed, would be built far away from the French planes, close to Leipzig, near a little town called Leuna.

BOSCH AND BASF were already thinking about the coming peace. Demand for sodium nitrate for explosives would go down, which put the investment in ammonia-to-Chilean salt-peter equipment at risk. Ammonia would again become the central product. But the Leuna plant was becoming so huge, capable of making so much ammonia that, along with the production from Oppau, there was worry that prices would drop, that the company would be drowning in ammonia when the war was over.

But perhaps that was not such a bad thing. Perhaps, using the government-subsidized plant, fixed-nitrogen prices in general would plummet so low that BASF would be able to drive its competitors, from the cyanamide makers to the Chileans, out of the market. Perhaps Haber-Bosch ammonia and the products made from it (like BASF's ammonium sulfate fertilizer) would become so cheap that they would dominate the world market. Since BASF owned all the patents, anyone else who wanted in to the market would first have to deal with the Germans. There would be licensing income from around the world.

At the very least, Leuna would provide Bosch's teams the chance to put into practice everything they had learned at Oppau. They would be able to perfect all the on-the-fly fixes, to integrate their biggest, newest ovens, to optimize production. Everything would be bigger. Everything would run more smoothly. To Bosch, Oppau was beginning to look like a test run. Leuna would be his masterpiece.

From the German military's perspective, the plant was both an immediate necessity and a sign of the future. The plant could be used two ways: fertilizer in peacetime, explosives in war. Both were important, especially if Chile remained out of reach. The easy switch from farming to arming could make Leuna into something like a secret weapon, a factory able to feed Germans during peacetime and protect them in war. As BASF had put it during the negotiations for the Saltpeter Promise, the directors hoped for "a permanent arrangement extending beyond the war, which would make it possible for us to supply the military for years to come." With Leuna, Germany would always be able to make its own bombs, its own gunpowder, and its own wars.

But for the moment it remained a balancing act between explosives and fertilizers: The more fixed nitrogen that went to bombs, the less there was for farmers. During the war, of course, the military was given priority. But as the fighting dragged on into the spring and summer of 1915, German farmers began to complain about shortages of fertilizer and their fears of crop failures. The farmers' pleas gave added justification to the government's massive investment in Haber-Bosch. The white salt made from ammonia could be used as fertilizer as well as the raw material for TNT; if Haber-Bosch could be expanded even more, it could be seen as an investment in agriculture as well as warfare. This added to the push for Leuna.

As Bosch threw himself and his teams into the Leuna project, Haber was starting something new. Victory was his only goal. He had helped solve the explosives shortage for Germany, but he was not confident that explosives alone were going to win the war. He turned his mind to developing another weapon, a weapon that did not rely on explosives, a weapon so devastating that just the thought of it would terrorize the enemy.

III

SYN

CHAPTER 13

FRITZ HABER WAS devoted to his kaiser, eager for glory, and full of ideas for ways that science could help Germany win the war. He hired a tailor and was fitted for a uniform that he helped design himself. The result was elegant enough, as a friend noted, to cause "quite a stir at headquarters." He became more Prussian by the day: organized, tireless and efficient, head shaved, mustache neatly trimmed, clothes immaculate, monocle at the ready. During this period of his life, Haber's existence revolved around, one of his sons later wrote, "an uncritical acceptance of the State's wisdom."

The same was true for many Germans. Germany was still a young nation, born in its modern form in 1871, close to the time Haber and Carl Bosch were born. The nation seemed, during the days before World War I, to mix adolescent insecurity and self-centeredness, vanity and thin skin, a hunger for respect with a quick temper. Germany's case was complicated by having a pair of older, more powerful siblings, France and England, who seemed to get all the attention. Germany had arrived on the world scene too late to build a big colonial empire and focused instead on building its strength internally, developing its industrial technology, its educational system, its military, and its science.

At the same time, Germans had a romantic notion of their

culture. They might be stereotyped as grasping, uncouth burghers by the rest of the world, but they saw themselves as *das Land der Dichter und Denker* (the land of poets and thinkers), the home of Goethe and Schiller, Beethoven and Bach, Kant and Nietzsche. Germany was relatively poor in land and natural resources, but it was rich in national pride. That fact was used by its leaders—especially Germany's "Iron Chancellor," Otto von Bismarck—to help forge what had been a patchwork of competing kingdoms and city-states into a single nation. His nation making was aided by creating a constant sense of threat, of having to fight against a ring of potential enemies, marked by early wars with Austria and France. Germans were destined for greatness, but to survive they needed a strong military. To survive they needed a strong king. To survive they needed to obey.

Unfortunately, the man they obeyed, Kaiser Wilhelm II, was a madman. Or Europe's most brilliant leader. Or delusional. Or the most glorious emperor in the history of Germany. It depended on whom you talked to, and when. In any case, it was clear to everyone who knew him that the absolute monarch of Europe's most technologically advanced nation was, as England's Lord Salisbury put it, "not quite normal."

One moment Wilhelm impressed everyone with his vitality, imagination, and quick grasp of situations. The next he threw a tantrum or launched into a conspiracy-laced tirade. His left arm, damaged in childbirth, was withered and almost useless, a shortcoming he tried to disguise by carrying gloves or placing his hand on the hilt of a sword. He used his strong right arm to humiliate others, once poking and pinching a young German prince until he brought him to tears, another time whacking the king of Bulgaria on the rump during a royal reception. He loved playing pranks, hunting, and living in the rough company of military officers. He seemed to need constant distraction.

To keep him amused and to cool his temper, his inner circle did tricks. A courtier once dressed as a poodle and barked for

Wilhelm at a royal gathering. To distract Wilhelm during a political crisis, the head of his military cabinet once dressed in a feather hat and tutu, and then died of a heart attack while dancing before his king.

When not amused, the kaiser could be dangerous. Wilhelm, on one military cruise, casually insulted a minor officer. The young man, to everyone's horror, struck his emperor. He was taken below, given chance to take his own life, and did. The kaiser's outbursts became stranger and less predictable as time went on. His advisers began to worry. Foreign ambassadors began to gossip. The British foreign secretary wrote his government (headed by a closely related monarch—Queen Victoria was Wilhelm's grandmother) that "Willy," as he was called in Buckingham Palace, was like "a battleship with steam up and screws going, but with no rudder." This was the man to whom a nation looked for guidance.

By the 1890s there was already talk of a coup to topple the erratic king (Wilhelm had come to power in 1888), but nothing ever gelled. The kaiser was smart enough to strengthen his position, stripping the aging Bismarck of the chancellorship, surrounding himself with acolytes, and restoring to the German monarchy some of the power that it had lost to elected officials. Although highly placed government officials worried, Wilhelm's show of strength appealed to his people. He seemed to believe himself a divine king and played the part. The more he rattled his sword and demanded international recognition, the more Germans seemed to love him. Even his critics finally accepted that it might be better to have a strong leader than a stable one.

When Wilhelm's policies and whims helped lead his nation to war, most of his subjects followed, singing. Germany was about to prove its greatness yet again.

• • •

THEY LATER CALLED World War I the chemist's war. That was certainly accurate for Germany, which was a technocratic, highly educated nation as well as a monarchy, the birthplace of Albert Einstein and Max Planck (two fathers of modern physics) as well as Wilhelm II. Despite the kaiser's power, Germany was anything but a monolithic state. Officials elected to the Reichstag, Germany's parliament, included liberals and socialists and labor activists as well as conservatives and monarchists. Debates were lively, a variety of views from aristocratic authoritarianism to socialist utopianism were tolerated, and science was honored by everyone. Science, it seemed, rose above politics, respected on one side for being once remarkably open and egalitarian while at the same time contributing in vital ways to industry and the needs of the state. Science created wealth and gave Germany power. True, the men who succeeded in reaching the highest position in German science were often those who most closely identified with Germany's national aims, but science also honored men (and even some women) of all political stripes and ethnic backgrounds, even Jews like Haber and Einstein, and allowed them to rise on their merits. Even Kaiser Wilhelm was a science enthusiast. He was proud of having his name on the Kaiser Wilhelm Institutes. He was even proud of scientists like Haber, a Jew who was also a Prussian-style patriot, and a highly respected scientist who had high-level contacts in business.

Haber worked hard to earn his kaiser's approval. Within weeks of the start of hostilities Haber was shuttling busily between his institute and laboratory, military headquarters, government offices, and industrial boardrooms. Any doors that had remained shut began to open to him. He became a top adviser, a sort of high privy counselor, a *Geheimrat* as the Germans called it, respected and welcomed everywhere in Berlin. The war gave him an outlet for his boundless energies, offered him a stage, and provided his restless mind with direction and meaning. He threw himself into it wholeheartedly. He transformed his institute into something close to a dedicated military research center. He outlined needs, brainstormed policies, arranged deals, made contacts, and wore his uniform whenever possible. He was a one-man military-industrial complex.

If there was ever a misconception that scientists are peaceable, Haber dispelled it. He was cold-blooded in his pursuit of victory. "Haber's actions continued to contradict Montesquieu's belief that knowledge makes men gentle," a fellow scientist wrote. "His boundless ambition seems to have made him determined to win the war single-handed."

He had another motivation as well. Despite his conversion, he was still considered a Jew by many people. Germany's Jewish population was, in the main, extremely enthusiastic about the war not only for the same reasons as other Germans, but because there was a sense that by fighting for their kaiser, they would prove their worthiness as Germans. "At least in war we are equal," wrote one German Jewish volunteer, shortly before he was killed.

THE MORE WARLIKE Haber became, the more his wife despaired. Clara Immerwahr, like Fritz, was born in Breslau to a successful Jewish family; Clara, like Fritz, was ambitious, bright, and passionate about chemistry. The difference between them was that Fritz's passion was given full play while Clara's was buried. They first met when she was a teenager and were reintroduced while she was studying chemistry at the university at Breslau. Clara was bright and charming, and Haber was beginning to feel as if his career was finally going somewhere. They were married a few months after she earned her degree, becoming the first woman at her university to earn a doctorate in chemistry. She was thirty years old.

Haber seemed very much in love with her at the beginning, but it soon became clear that his only real passions were his science and his nation. She was dissatisfied with their marriage even before she got pregnant. When their son Hermann was born in 1902, Clara felt a trap closing. The feeling was heightened when Haber left, a few weeks after the baby arrived, on a four-month tour of science facilities in the United States. The marriage never recovered. When he was home, he treated her distantly, sometimes, she thought, cruelly. She wrote the professor who guided her work at the university, "I'd rather write ten dissertations than suffer this way."

The problem was that Haber had married a talented Ph.D. but wanted a traditional German wife, happy to look good for him and to run a comfortable home, a woman devoted to *Kinder, Kuchen, Kirche* (kids, cooking, and church). He expected that after their marriage his wife would naturally put science behind her and throw herself happily into domestic life. Anything beyond that seemed to baffle him. "Women are like lovely butterflies to me," he told a friend. "I admire their colors and glitter, but I get no further."

Instead of becoming the perfect wife, Clara began to suffocate. Around the time her husband was perfecting his ammonia machine, she poured her heart out in a letter to a friend:

> *What Fritz has achieved in these eight years, I have lost—and even more. And what's left fills me with the deepest dissatisfaction. . . . [E]ven if external circumstances and my own particular temperament are partly to blame for this loss, what's mainly responsible, without a doubt, is Fritz's overwhelming assertion of his own place in the household and in the marriage. It simply destroys any personality that's incapable of asserting itself against him even more ruthlessly. And that's the case with me. And I ask myself if superior intelligence is enough to make one person more valuable than another, and whether aspects of myself that are going to hell because they haven't met the right man aren't more*

important than the most significant elements of electron theory. . . . Everybody has a right to live their own life, but to nurture one's "quirks" while exhibiting a supreme contempt for everyone else and the most common routines of life—I think that even a genius shouldn't be permitted such behavior, except on a desolate island.

The move to Berlin and Haber's new position as head of an institute only made things worse. Clara withdrew into herself, growing moody, melancholy, and anxious, while trying to maintain a facade of domestic responsibility. Those who knew her in those days remembered her turning into "a gray mouse, inconspicuous and nice," dressing in rough woolen clothes with a white apron, spending all her energy on her son, whom she "cared for and pampered in such a way," a relative wrote, "that we made jokes about it." She was loathe to leave the house. She was fading.

Fritz, on the other hand, was often at social events; he sparkled at lively dinners and parties. He often spent the evening at a Berlin private club, dressed in evening clothes and enjoying a drink while he chatted with the other members. At some point he was introduced to the club's business manager, Charlotte Nathan. She liked to talk, knew how to dress, and enjoyed socializing. She was quick-witted, lovely, Jewish, and very much younger than Haber. She was as much a butterfly as Clara was a mouse. Haber began spending more evenings at the club.

A FEW MONTHS into the war, Haber's institute had grown and changed. It was now rich with government funding, staffed with hundreds of new workers—and surrounded by barbed wire. Soldiers kept guard at the entrance. Inside, he and his assistants were working on Haber's secret weapon. He had been thinking about trench warfare and believed that an enemy who was well dug in was almost invincible. The war would not be won with bombs and massed attacks, he thought. Victory would require something very different. Something terrifying.

"Every war is a war against the soul of the soldier, not the body," Haber said. "New weapons break his morale because they are something new, something he has not experienced, and therefore something that he fears. We were used to shell fire. The artillery did not do much harm to morale."

The weapon he sought was found in a test tube. It was a weapon that would crawl silently along the ground, creep into the trenches, and suffocate its victims. A weapon that would destroy spirit as well as body. A weapon that could not be stopped.

The idea of gas warfare was not new. It had been around long enough to have been banned, with Germany and many other nations pledging in 1899 not to use poison of any sort in warfare, including "projectiles the sole object of which is the diffusion of asphyxiating or deleterious gases."

The French were the first to break the pledge, lobbing shells containing a sort of primitive tear gas at the Germans early in World War I. The intent was to drive Germans from their positions rather than kill them. The Germans responded with their own version of a gas irritant, mounting an attack so ineffective that the Allies never even noticed it. The British too started researching ways to make gas bombs. But nothing anyone developed had been much of a success.

Haber decided that he could make gas into a superweapon, one that could quickly win the war for Germany. His enthusiasm led him to an appointment as head of a new chemicals section in the Prussian war ministry, a position that earned him the much-desired rank of captain (something of an honorary position, outside the usual officer training regimen and therefore open to a Jew).

Captain Haber began working on his superweapon. There

were several problems to solve. Most gases quickly dispersed in the air, rendering them relatively harmless. He wanted a heavier gas that would advance like ground fog, surging into trenches as it advanced. Artillery would not be an ideal method of delivery because it would disperse the gas too much. Haber began thinking of a line of large canisters set along the front, each filled with a heavier-than-air gas. Opening them at the same time would release a thick cloud, which, if the wind was right, would move toward the enemy, choking them, driving them back. Haber did not want a tear gas. He wanted a killer gas. Only the threat of death would destroy morale, spur panic, and trigger a rout. It would all happen in eerie silence. There would be no defense. The enemy would throw down its weapons. It worked legalistically as well, because the use of canisters along the front also would get around the 1899 pledge against the use of gas projectiles. It was a technicality but one that seemed important to the Germans: There would be no projectiles.

Haber's institute considered many gases before settling on what appeared to be a perfect candidate: chlorine. Pale green chlorine gas was slightly heavier than air and highly toxic. There was plenty of chlorine around because it was an industrial raw material used for making dyes. BASF made it in bulk and, with the dye market drying up because of the war, had chlorine-making machinery sitting idle. So did other dye companies. It would be easy to make.

Haber perfected his idea through the first months of 1915. He organized an elite company of engineers and scientists, including three future Nobel Prize winners (Gustav Hertz, Otto Hahn, and James Franck), to work on it. He came up with a plan for deploying canisters along a fifteen-mile stretch of the front, tended by specially trained troops who would release the gas in precise synchrony when the wind was right. The demoralized enemy would break and run. Behind the toxic cloud would march thousands of German soldiers wearing gas masks, sweeping through the broken enemy lines to victory. Hardly a shot would be fired. Haber's gas would open the door to Paris and victory. When one of his colleagues expressed reluctance about gas warfare's seeming violation of international accords, Haber reminded him that the French had already used gas. In any case, he argued, if the chlorine attack worked, the war would end more quickly and "countless lives would be saved."

As the project moved to field tests, one of Haber's men remembered him striding across the testing area in his uniform, "cool, unafraid, and death-defying in the most advanced positions." He seemed, finally, to be in his element. Clara accompanied him to one test, anxiously watching her husband's weapon in action, and a trooper remembered her "a nervous lady who was sharply opposed to his accompanying the new gas troops to the front." The dangers were real; during one test, Haber rode his horse too close to a gas cloud and nearly suffocated.

Haber perfected his gas system quickly, but the German military commanders were not as eager to deploy it. The whole idea was unproven. The concept of gas warfare was, to many officers, distasteful. One German commander wrote his wife, "I fear it will produce a tremendous scandal in the world," that in response the Allies would "soon have something similarly diabolical."

"War has nothing to do with chivalry any more," he wrote. "The higher civilization rises, the viler man becomes."

Personal feelings aside, the German commanders worried about practicality: Haber's secret weapon would take too long to set up; the enemy would see what they were up to and the element of surprise would be lost. But all objections were set aside as the war's body count rose. Finally, as the murderous battle of Ypres dragged on, permission was granted to mount a gas attack in the area, on a reduced scale, over a shorter length of the front. They called it Operation Disinfection.

In March 1915 Haber's gas troops began placing their canisters along the lines at Ypres, burying them in the soil under cover of night to avoid tipping off the enemy. They were working near the sea in a less-than-ideal spot—the wind most often blew in the wrong way, off the ocean, into the faces of the Germans—but it was a start. Haber hid his canisters and waited for the wind to shift. For weeks gas attacks were planned and men gathered, then dismissed as the weather conditions refused to cooperate. No wind was just as bad as a contrary wind. The German commanders began to think that it was all a waste of time; the German soldiers grumbled as attacks were called on and off. Enemy bullets or shrapnel occasionally punctured a hidden tank and released chlorine, injuring the kaiser's troops; the danger of the whole scheme's being discovered increased as time went on. While the attack was delayed, complaints started to increase. Crown Prince Rupprecht of Bavaria, who commanded a German force dug in near the canisters, complained that the whole thing made no sense because the Allies would simply copy the gas idea, use it against the Germans, and have more success because in this section of the front the wind was at the Allies' backs. The German high command began to lose faith in the scheme.

Haber waited nervously, arguing that it would take months for Germany's enemies to retaliate in kind, that their chemists were far behind Germany's both in preparation for this kind of war and perhaps in talent. By the time they were ready to use their own gas, the war, with any luck, would be over.

By mid-April more than fifty-five hundred of Haber's gas canisters had been placed in the German forward trenches, and the decisive attack was scheduled for the twentieth. But the delays had cost Haber some of his support. German commanders ordered away many of the reserve troops who were supposed to surge forward behind the gas, sending them to fill needs on the eastern front. At the highest levels of planning Haber's grand, unproven scheme for a gas attack was now seen as something more of a diversion, a novelty they could wave in front of the Allies to screen other screen-troop movements. If the gas gained them a bit of ground, fine. But this business of a Jewish scientist's planning military strategy was not to be taken too seriously. Then the attack planned for the twentieth was called off yet again, as the wind continued to blow from the sea.

The offensive was reset for the morning of April 22, 1915. Thousands of German troops were again gathered. And again the wind refused to cooperate. Through the morning and into the afternoon it kept blowing toward the German lines. Late in the day it finally reversed, and at long last the order came to release the gas. About 5:00 p.m., with perfect precision, Haber's gas troops opened their canisters. A wall of chlorine gas four miles long began slowly advancing toward the French.

It worked just as Haber had predicted. The gas caused a panic. A British soldier stationed several miles from the attack remembered the scene, with a strange cloud of yellow-gray gas moving forward, a strange tang in the air, and then, "Suddenly down the road from the Yser Canal came a galloping team of horses, the riders goading their mounts in a frenzied way; then another and another, till the road became a seething mass with a pall of dust over all." The horses were throwing froth, the riders wild-eyed, sometimes two or three to a horse. They were followed by thousands of troops on foot, many of them French colonial soldiers from Algeria, mobs of men running as fast and as far as they could, throwing away their rifles and packs, even their tunics.

As Haber had hoped, the attack opened a huge gap in the Allied lines. But the Germans squandered their opportunity. German soldiers, unsure about the remaining gas, moved forward tentatively. For an hour in the dusk they probed their way into an eerily deserted landscape. There was no major resistance. But they were too slow. As they felt their way forward, Canadians stationed to the sides of the gas attack held their line and began pouring flanking fire into the advancing Germans. As darkness fell, the Germans stopped. The reserve troops needed to complete the breakthrough had been sent away. That night the Allies marched in more troops to seal the gap. Although the Germans conquered a sizable chunk of the area, they lost their chance to win the war.

However, the Germans had made a significant advance through entrenched positions with very low casualties. In fact, there had been remarkably low casualties among the Allies as well; a German count of the enemy dead the next day found that the gas attack had killed no more than a few hundred soldiers. Again, the gas had worked as Haber had hoped: Its effects had scared the enemy away rather than causing mass death. The news caused a stir at German general headquarters, and the kaiser himself was ecstatic when he heard about it, embracing his army commander three times and ordering a bottle of champagne for the colonel in charge of that section of the front. Haber was asked to appear, finally, before his emperor, and was awarded an Iron Cross.

The celebration was short-lived. The Allies recovered from their surprise, educated their troops, and gave their men improvised respirators made from wetted handkerchiefs. When the Germans tried another, smaller, chlorine release two days later, the Canadians under attack were pushed back a bit but not crushed. Then the Allies began planning their own gas attacks. The first British attempt made at Loos five months later, using chlorine released from canisters, failed when the wind shifted, blowing gas back into their faces. It is estimated that as many British soldiers as Germans died that day from the gas. Then both sides went on to develop more, better, deadlier poison gases.

CLARA HABER, WAITING at home, told a friend that she was in despair. The news from the front was all too horrible, chemicals being used as weapons. She had been shocked just a few months before when a friend working at Haber's institute had been killed in a chemical explosion, a lab accident, and now her husband was masterminding the murder of thousands. She loved chemistry. This was a perversion.

After his limited success at Ypres, Haber returned to Berlin, where he was fêted, honored, and ordered to the eastern front to prepare gas attacks against the Russians. Before he left, Haber threw a party at his home. It was said later that the guests included Charlotte Nathan, the vivacious young business manager of Haber's Berlin club.

Later that night, Clara, according to her biographer, spent hours writing several letters. Then she went to the villa's garden, carrying her husband's service revolver. Her son Hermann, then twelve years old, heard two shots. He ran to the garden and found her. She had aimed for her heart and was bleeding badly, but she was still alive. Hermann called his father, who had, it was said, taken sleeping pills. By the time Fritz got to her it was too late.

It was assumed that the first of the two shots was Clara testing the gun. None of her final letters survived.

Haber, a few days later, boarded a train for the eastern front.

IT IS EASY to think of Clara's suicide as an act of political protest. Haber's friend James Franck, who helped with the gas work, later said that Clara "was a good human being who wanted to reform the world. The fact that her husband was involved in gas warfare certainly played a role in her suicide." But she was also disappointed in her life. "It was always my view of life that it was only worth living if one developed all one's abilities to reach the heights and experienced as much as possible of what a human life can offer," she had written years before. Clearly she felt stifled as a *Hausfrau*. She no doubt envied Haber his brilliant career. She appears to have been prone to depression; others in her family, including a sister, committed suicide. She was in despair for a number of reasons, not just because her husband was involved in gas warfare.

Haber's response to her death, his quick departure, his abandonment of their son was complex too. It is likely that he destroyed or hid Clara's suicide notes. But, as the historian Fritz Stern contended, "The oft-repeated assertion that Haber reacted coldly or indifferently to his wife's death is erroneous." He left for the eastern front because he was ordered to and because allaying the emotional pain with work was the only thing he knew how to do. Haber did not understand his wife, which is not the same as saying that he did not love her. "For a month I doubted that I would hold out," he wrote later from the front, "but now the war with its gruesome pictures and its continuous demands on all my strength has calmed me."

HABER'S ENTHUSIASM FOR Germany's cause was so great that in October 1914 he joined more than ninety other German intellectuals in signing an ill-advised public statement called the Manifesto to the Civilized World. It read in part: "As representatives of German Science and Art, we hereby protest to the civilized world against the lies and calumnies with which our enemies are endeavoring to stain the honor of Germany in her hard struggle for existence—in a struggle that has been forced on her. . . . Were it not for German militarism, German civilization would long since have been extirpated. . . . Have faith in us! Believe that we shall carry on this war to the end as a civilized nation, to whom the legacy of a Goethe, a Beethoven, and a Kant, is just as sacred as its own hearths and homes." Other signers included current or future Nobel Prize winners Max Planck, Wilhelm Ostwald, Paul Ehrlich, and Walther Nernst.

It backfired. The manifesto was roundly attacked in other

nations as a mark of how jingoistic even the best minds in Germany had become. It succeeded only in fanning anti-German sentiment. Haber, thanks to his signature, was established in the mind of the Allies as less a freethinking scientist than a slavish apologist. When it became known later that he had masterminded the gas attack at Ypres, he became more than that. He became, in the minds of many, a war criminal.

HABER'S WORK WITH fixed nitrogen had brought him to Berlin and given him the chance to develop chlorine gas. Fixed nitrogen would shape Carl Bosch's life as well. As Haber waged war, Bosch focused on building a giant factory planned for Leuna, designed to be the biggest, most innovative industrial facility the world had ever seen, a single machine the size of a small city, rising fresh from raw farmland. It was made possible by the military's need for explosives. The deal BASF had signed with the government in April 1916 pledged the company to achieve, within about a year, production of five thousand tons of nitrates per month, rising to seventy-five hundred tons by August 1917. It was a brutal time line. But the war was not going well for Germany, which remained bogged down in France while its allies, notably the Ottoman and Austro-Hungarian empires, weakened. The Army was burning explosives at an unheard-of rate. Bosch's new plant was absolutely necessary if the war was to continue.

The BASF teams began working feverishly, tying together the many systems needed to feed gases into the ovens, the huge oxidation outfits, the new sections designed to transform ammonia into nitrates, bigger heat exchangers, improved feedstock gas production units, railway systems in and out, water systems, power systems, worker's quarters, worker's transportation, storage facilities, shipping facilities, administrative support facilities, a thousand subprojects, a million details.

Two and a half square miles of pastures and fields, purchased from unhappy local farmers at confiscatory prices (sales
enforced by military orders), was leveled and turned into a huge
construction site occupied by a small army of construction
workers. Many were skilled BASF workers but not as many as
Bosch wanted; almost half of the company's 1914 workforce
had been conscripted into the army. To build Leuna the lost
workers were replaced with soldiers, prisoners of war, hundreds of Belgian civilians and forced laborers, and—for the first
time in the company's history—women. Everyone who could
hoist a tool, it seemed, was shipped to the area. A temporary
barracks-city was built to house them.

What they accomplished amazed even Bosch. Seven months
after breaking ground in May, steel skeletons were being turned
into giant buildings ready to be fitted with machines. Bosch
oversaw everything personally, but operational details were carried out by a group of top assistants. Bosch came to rely especially on Carl Krauch, the young chemical engineer who had
figured out how to stop a corrosion problem at Oppau. Krauch
became Bosch's main problem solver. When bad weather and
iron shortages threatened the project in early 1917, Krauch figured out how to get things moving. Krauch kept crews working around the clock. Krauch made sure they hit their deadlines.

On April 27, 1917, the first ammonia oven was fired up. The
next morning, the first railroad tank car was filled with Leuna
ammonia. On its side, a worker scrawled "Frenchmen's death."
The plant was built not to produce ammonia, however—at
least not primarily—but nitrate, the raw material for explosives. This it started doing a little behind schedule, but not
much. The great plant took less than a year to go from raw land
to a fully functional production facility. Output rose fast, from
36,000 tons per year at opening to 160,000 tons per year by the
end of the war, with plans in place for a grand expansion to a capacity of 240,000 tons. There were occasional breakdowns and

repairs, of course; just as at Oppau, every part of Leuna had to feed smoothly into the next, and a breakdown in any one part could stop the entire process. But Bosch's giant new plant was so well designed, so carefully managed, that errors were fixed quickly.

Almost all Leuna's output was fed directly into the German war machine. The importance of the company's contribution was recognized in July 1917 by Field Marshal Paul von Hindenburg himself, Germany's supreme military commander, who sent the directors of BASF a much-prized autographed photo. "The times are hard," it was inscribed, "but victory is certain."

LEUNA TRANSFORMED BASF from a dye company to a nitrogen company. It proved Bosch's abilities and led to his rapid promotion within the company. Its enormous output shouldered other, competing nitrogen-fixing technologies, cyanamide and the arc process, out of the market. Finally, it put BASF firmly into bed with the government.

By 1918, running at full capacity, Leuna had become Germany's industrial marvel. It was bigger than any Ford plant. It used technology that no one else could duplicate. It kept Germany in the war. Some historians have estimated that World War I would have ended a year, perhaps two years sooner, if Haber-Bosch had not been able to make the nitrates needed for explosives.

The fact that Germany did not win the war had little to do with technology and much to do with the tightening of the British sea blockade and the addition of the United States to its list of enemies. The arrival of the Yankees in France outweighed everything German scientists had made possible. After one last great German offensive failed in the spring of 1918, its allies Bulgaria, the Ottoman Empire, and the Austro-Hungarian

Empire began giving up. By the fall of the year they were suing for peace. War-weary German sailors, soldiers, laborers, leftists, and antimonarchists started uprisings that flared into a revolution. Finally, the kaiser's generals estimated that the Allies were within days of breaking the German lines in France. When it became apparent that the war was lost, the kaiser abdicated and fled to Holland. On November 9, 1918, without an enemy ever crossing its border, Germany surrendered. One-tenth of its prewar population was dead.

CHAPTER 14

L OSING THE WAR was like losing the top to a pressure
cooker: Everything boiled over. For more than a year
after the kaiser's abdication, Germany collapsed into
something close to anarchy. The spontaneous workers' and soldiers' revolts around Germany in the final days of the war co-
alesced in cases into full-scale revolution. A soviet republic was
briefly declared in Berlin. The leftist uprising was quelled with
the help of the Freikorps, a right-wing paramilitary force made
up of former soldiers. The Freikorps in turn became part of its
own revolution, from the Right, which in turn was suppressed
by forces in the center. In the clamor and confusion a shaky government emerged, an attempt at a centrist democratic republic.
It was named after the town where it was declared: Weimar.

From the start, Germany's Weimar Republic faced overwhelming challenges, pressured by both left and right, always a
bit too weak and tentative to earn full popular support. It faced
a world that seemed intent on continued punishment. Anti-
German sentiment among the victors was intense. German-
owned factories in the United States and other Allied nations
were shuttered or sold at auction and their industrial secrets
seized. German chemical patents were an especially tempting
target now that the Great War had shown just how far behind
Germany most nations were.

This was a critical moment for BASF. When the war ended, the victors occupied what had been German land all the way to the Rhine, including the area around the ammonia factory at Oppau and the original dye works at Ludwigshafen. Bosch knew that it would be difficult for competitors to replicate his success with fixed nitrogen on the basis of seized patents alone. But it would be different if they seized the factories, reviewed their operation, and dissected an oven or two. If that happened, there was a very good chance that the ammonia secret would be out, and BASF would lose all its competitive advantage. Even in the midst of political chaos, business was business.

THE FRENCH OCCUPATION would stop at the Rhine. The problem was that the BASF factories, with their valuable stocks of chemicals, were on the west side—the French side—of the river, an area of the Rhineland that had been German territory for decades, but which the French were now arguing should become a demilitarized buffer zone between the two ancient enemies. The important thing to Carl Bosch was that two of his company's three main factories would be under French control for who knew how long. At least Leuna was safe, far inside Germany, away from the occupation. Before the French marched in on December 6, 1918, BASF carted as much of their inventory as possible across the bridges of the Rhine to safety. The German company had time to move a great deal of its stock. But not all. The French confiscated what remained— BASF estimated it at more than a million marks worth of finished dyes—and shipped it to France.

Before the French arrived at Oppau, Bosch shut the ammonia ovens down, intending to keep them cold until the occupying force left. His stated reason was that there was a shortage of coal for powering the plant, which was true. The real reason, however, was that he did not want the French to see his ovens in action. When the French demanded that he bring Oppau back online, he and the BASF lawyers argued that they had the right to keep them idle, because the French were clearly after industrial secrets about fertilizer production. As the Germans read the articles guiding the peace, the occupying forces had a right to demand answers to questions about industrial raw materials and products but not about the processes used to turn one into the other. Just because the Allies won the war did not mean they had the right to steal German technology.

Or did it? The French seemed to believe otherwise, as did the British when their inspection team arrived at Oppau. They considered Oppau to be a war munitions plant, not a fertilizer plant. They were right about the first part and wrong about the second. The plant did both things. While the occupying forces proceeded on the understanding that they had a mandate to examine and dismantle the German war machine—including the bits that supplied vital materials to the explosives industry—the Germans held them back by arguing the proprietary rights of peacetime.

Bosch refused to give them free access, using every legal maneuver he could to stall the occupiers while he waited for the situation to resolve itself. Much would depend on the outcome of the formal peace talks, being scheduled for Versailles later in 1919. The French, meanwhile, intent on demonstrating their power, set up a military post at Oppau; demanded accounts of stocks of chemicals, supplies, and raw materials; and insisted on reviewing all documents related to production methods. They photographed the BASF workers as if they were prisoners, scores at a time, with their heads sticking through a wooden latticework so that each face could be separately identified. French inspectors, some in long coats and bowler hats, some in military uniforms, poked through every factory building. A number of them were chemists and engineers employed by private companies. They carefully measured every machine at Oppau and

Ludwigshafen, gathered samples, tested materials, interviewed staff, climbed ladders, peered into pipes and vats, their fingers stained violet with dye, "creeping about everywhere," as one BASF employee put it.

Whenever a French inspector appeared, the BASF employees laid down their tools. Any machine in operation (the company had stopped making ammonia but continued making dyes) was switched off. They stared at the French until they left. The inspectors would arrive at a building to find needed ladders mysteriously missing and, in at least one case, an entire flight of stairs. Important gauges sometimes disappeared or had their faces blacked out.

Without seeing the ammonia ovens in operation, the French could not figure out the Haber-Bosch system. Bosch and his people knew that the technology was too complex, too precise, and too interconnected for easy comprehension. Visiting French inspectors were told before entering the gates of Oppau that "if they saw the plant they could not be able to reproduce it, and even if they erected it, they would not be able to work it." Whenever the French complained about being hindered in their duty, BASF countered with its own complaints about French soldiers' abusing local girls, smoking in dangerous areas of the plant, or playing soccer in Bosch's agricultural test gardens.

Bosch was playing a difficult game. In the long term, he needed to deny his ammonia-making secret to the Allies. In the short term, however, he needed to make money for BASF and keep his workers busy. He did not want his people to start listening to communist organizers. Since they were not making ammonia at Oppau, he set thousands of employees to work on repairs and new construction, fixing and overhauling sections of the plant, getting it in top condition for the day the French left. That too was an overriding issue. No one knew how long the French were going to occupy the area—there was talk that it could go on for years, perhaps decades. In the meantime he kept his workers on the payroll despite the fact that the company was losing tens of thousands of marks every week from the Oppau shutdown—more than twenty-five million marks in the first four months of 1919 alone, according to one estimate. But it was better than losing the technology itself.

That was a very real threat. Everyone now understood that Haber-Bosch had kept Germany in the war, had armed its soldiers and fed its people despite a complete blockade of shipments from Chile. Every nation wanted its own plants. In the United States, Congress had voted millions of dollars to build a Haber-Bosch plant during the war, but the attempt had been a dismal failure. So had a try in Britain. The technology was simply too complex. The fastest way to break the secret would be to pull apart the plant at Oppau and study it until the method was clear.

Where the French could not succeed, the British intended to. A team of British inspectors arrived at Oppau in the spring of 1919, including researchers from the big British chemical firm Brunner Mond. They had never seen anything of the size and scale of Oppau (much less Leuna, which was far away and operating in unoccupied Germany). "Wonderful," one of the British chemists wrote after his first day of inspection. The British, too, were told by the Germans that the ammonia equipment was inoperable; their request to closely examine the ovens was refused, and attempts to have parts of the plant dismantled for examination were politely and firmly rebuffed. The British tried to get the French occupying force to back up their demands. When the French approached BASF, they were told that anything beyond a cursory British examination would cause the company to shut the plant down entirely, throwing thousands of local residents out of work. The French could then deal with the consequences.

In the end, the disappointed British were left with only the notes and sketches they had made. These were put in a locked railway wagon, under armed guard, before they were to leave. That night someone managed to get underneath the wagon, cut the bottom out, and steal everything. The British team had to make its official report from memory.

BOSCH CONTINUED TO stall for time. He was waiting for the situation to settle, for the new German government to grow strong enough to help his company, for final rules to be laid down regarding the occupation and the rights of the Germans. He pinned his hopes on the official peace negotiations, now set for March 1919.

Bosch was named a representative of German industry to the talks at Versailles. Upon his arrival he was, like all the German delegates, thrown into prison. It was not called a prison—it was called protective custody—but to the Germans it was a high-class prison nonetheless, a hotel surrounded by barbed wire and armed guards. Entrances and exits were watched and noted. A curfew was imposed. It was a slap in the face of the German delegation, a clear demonstration that they were not to be treated as equals.

That fact was quickly reinforced when the talks began. The Germans quickly realized that they were there not to negotiate a peace but to accept one. There was little incentive for the Allies to bargain. Both sides had lost unthinkable numbers of men. But while the war had devastated large areas of France and Belgium, German territory, German factories, and German homes had escaped relatively unscathed. Now it was time to pay.

The French, who had in many ways suffered most, were the most adamant about bringing their longtime enemy to its knees and destroying forever its ability to make war. They were backed by strong anti-German public opinion fueled by worldwide wartime propaganda and the daily reminder of armless, legless, and blinded veterans on the streets of every town in Britain and France, and many in the United States, too. The French demanded the destruction of all German armaments plants, not just those that made gunpowder, weapons, and explosives, but also those that made the raw materials to make explosives and poison gas. The French wanted to shut down Oppau and Leuna.

This was extreme although not unexpected. More worrisome was the talk of vast reparations payments, money for the victors to cover the damage they suffered, which were to be paid not only with German gold but also with dyes, ammonia, and anything else of value. The amounts being discussed were astronomical, high enough to ensure that Germany would remain economically crippled for decades.

Bosch made the case that his dye and nitrate plants were needed by the Allies, not just to help feed the German people during this difficult time but in addition to prevent German disorder, keep the peace, and employ Germans. Germans at work were Germans less likely to be lured into revolution by the arguments of the Bolsheviks. If the Allies wanted to contain the spread of Russian-style revolution, they had better start in Germany. Plus, if the plants were closed, how was Germany to earn the money for its reparations payments? His arguments were listened to politely, then, for the most part, ignored.

Something else was needed if he was to keep his plants from being closed or destroyed. Under cover of darkness late one night, Bosch climbed over the wall and wire surrounding the hotel, avoided the guards, and made his way through the streets of Versailles to a secret meeting. He spoke there with a highly placed representative of the French chemical industry, businessman to businessman. It was straightforward. Bosch had only one thing to offer, and he offered it to the French: the chance to build a Haber-Bosch plant. In exchange he wanted a promise that Oppau and Leuna would stay open. There was a bit of money involved too. The French government would be given exclusive rights to the Haber-Bosch processes within French territory. A French national corporation would build an ammonia plant to BASF's specifications, capable of producing one hundred tons a day, and would have access to any technological improvements discovered by BASF for the next fifteen years. In exchange the French would pay BASF five million francs (considerably lower than the fifty million Bosch had initially asked for) plus a small royalty on every ton of ammonia produced in their plant.

The French would get what they wanted—an operating Haber-Bosch plant—and Bosch would keep his nitrogen plants running. The papers were completed on the first anniversary of the Armistice. The French inspections of Oppau ceased within a few weeks. The occupying French forces pulled out in early 1920. The ovens were fired up, and Oppau slowly began working back toward full production.

Bosch had sidestepped one disaster. But there were more.

BOSCH WAS REWARDED for his negotiating skills by being made head of the company, something he both wanted and dreaded. He lived and breathed BASF, thought himself skilled enough to run it, but hated the prospect of endless meetings. He would rather spend his time tinkering with machines than figuring out personnel issues. But his sense of mission was engaged now. He wanted to guide BASF's future in a harsh postwar world. Of course the salary and perks (including travel and a private villa) were welcome, to both him and his wife.

The company was still in trouble. The French were selling the German dyes they had confiscated as reparations, flooding the market and lowering prices. At the same time, dye production was rising in former enemy nations as the United States, France, and Britain—sometimes using confiscated German patents—built their own factories. The Germans could do little about it.

The war had chipped away at Germany's preeminence in chemistry. The Allies had come to appreciate the role of scientists in wartime. With the flow of German chemicals and dyes cut off during the war, the United States and Britain had to invest in their home industries. When the war ended, they were well on their way to catching up with the Germans.

It all served to hasten the end of the old German dye industry. Bosch responded with two initiatives: Boost ammonia production, and find the next big idea. The Haber-Bosch process was stronger now than it had been at the beginning of the war, thanks to the construction of Leuna. It was still the most efficient way to fix nitrogen. The postwar world was hungry for food, which meant it was hungry for fertilizer. Dyes were out; Haber-Bosch was in.

At the same time, Bosch knew that one way or another, the Haber-Bosch system would eventually be duplicated in other countries. The deal he had been forced to make with the French merely hastened the process. Once there was an operating plant in France, no doubt the technology would spread. Bosch began the long process of looking for something as big as synthetic ammonia. Or bigger.

GERMANY WAS IN the process of having its teeth pulled. Through the early 1920s the Allied powers continued disarming the nation as outlined in the Treaty of Versailles, including the inspection of chemical plants to make sure they were not being used for military purposes. Bosch and the other leaders of German chemistry continued to provide careful, slow cooperation, filling out dozens of Allied questionnaires and allowing occasional inspections. Their tactic was to cooperate fully whenever it came to actual military stores—powder, explosives, and propellants—but to continue fighting any attempts by the Allies to learn about or interfere with the techniques used to make ammonia or fertilizers, products that the Germans argued were for peaceful uses. According to the letter of the Treaty of Versailles, the Allies had no right to interfere with or demand secrets about any industrial production unrelated to warmaking. However, as everyone knew, Haber-Bosch plants could easily be switched to making raw material for explosives. Public sentiment among Germany's former enemies in favor of shutting down the plants was increased in the 1920s by books like *The Riddle of the Rhine*, published in Britain and the United States, which pointed out that German disarmament could hardly be said to be complete as long as the German firms in any way involved in making gunpowder and explosives were still operating. There was a cry for total "chemical disarmament." Allied politicians used the threat of reinvading the Rhineland to force greater German cooperation. The Germans continued to give every indication of cooperating, while protecting their industrial secrets. The French, thanks to their deal with Bosch, were not as energetic about it as they might have been. Bosch held on, hoping for the day when the Allies would recognize that it was no longer profitable to keep punishing Germany.

Bosch was working very hard, overseeing the protection of his plants and negotiations with the occupying forces on one hand, trying to keep his workers happy and away from the Communists on the other; protecting the German dye making industry as long as he could while working to expand Leuna (finishing a final phase of growth that had been planned with the government during the last days of the war); and seeking the next technology, the next big breakthrough that would keep his company healthy twenty years down the road. He was under enormous stress.

Yet he did find moments of peace. One day an ornithologist from Heidelberg was bird-watching along the Rhine when he heard a soft splashing in the reeds. Creeping closer, the bird-watcher saw a man knee-deep in the water, his pants legs and shirtsleeves rolled up, dipping a net. The man pulled up a dripping load of muck and began feeling through it. He was hunting freshwater mussels. The two struck up a conversation. The mussel hunter was Bosch, taking a few hours off, trying to find a way to relax. He still loved hiking and collecting specimens of animals and plants. Sometimes he would have his driver take him up into the hills; when he could not get away that long, Bosch would make do with whatever piece of nature he could find. Here, along the banks of the Rhine, often alone, he could put the company aside and lose himself in the sound of the water and birds.

Fritz Haber was not so lucky.

CHAPTER 15

TWO YEARS AFTER Clara committed suicide, Fritz Haber asked Charlotte Nathan to marry him. It surprised no one and seemed to please no one. The former business manager of Haber's Berlin club was twenty years younger than her new husband and an "extraordinarily attractive woman," according to a Haber biographer. At Haber's insistence, Charlotte, like Clara, converted to Christianity, which allowed them to be married in Kaiser Wilhelm Memorial, a grand neo-Romanesque church in the center of Berlin. It was 1917 and the war was entering its fourth year. Haber wore his uniform, replete with dress sword and spiked helmet.

Ten months later, as the war finally ground to its end, they had a new child and a failing marriage. Soon after his daughter was born, Haber left home for a stay at a health spa. Charlotte began complaining to her friends about Haber's workaholism and self-obsession. Haber mused about his wife's demanding ways. Charlotte was not depressed or withdrawn like Clara. She was sharp, energetic, and spoke her mind. Haber found that the difference in temperament made no difference in his happiness. He understood his second new wife no better than he had his first.

By the time the war was over—his marriage troubled, his military efforts futile, and his nation in chaos—Haber was a psychological wreck. "He was overwhelmed by the outcome of the war," wrote an observer, "and, for several months, nervously exhausted." Fatigue was part of it. So was a general sense of disorientation, a loss of purpose. "You know the feeling when you're on a snow-covered slope, sliding downward?" Haber wrote a friend a few weeks after the war ended. "You don't know until you get to the bottom whether you'll arrive with all your limbs intact or with broken legs and neck. All you can do during the slide is stay calm." That was how life was in Germany. He packed away his uniform and watched as Germany reeled from one crisis to the next, through the disbanding of the German army, the Bolshevik-inspired revolts, the insults of Versailles, and the spread of poverty, hunger, and disease.

There was to be one more insult, as personal as it was national. In the summer of 1919 Haber got word that his name appeared on a list of German war criminals. French and British authorities, he was told, were going after him because of his work on chemical weapons. If he stayed in Germany he risked arrest and trial. Haber gathered his wife, his seventeen-year-old son Hermann, and his new daughter, bought a forged passport, and fled to Switzerland. He had money enough to speed the process of obtaining Swiss citizenship. He and his family settled in St. Moritz. He grew a beard.

After a few months the talk about capturing war criminals died down. It turned out that the rumors had been inflated: Haber's name did not appear on any arrest list. It was safe for him to return to Berlin and his institute, but he would carry with him a dual reputation: world-respected chemist and infamous gas-attack mastermind. As the passions of the war cooled, his first, larger reputation took prominence. Shortly after coming back, in November 1919, his fortunes swung again, this time to the positive side, with news as welcome as it was surprising. He was named the winner of a Nobel Prize in chemistry in honor of his synthesis of ammonia. Haber's wartime infamy and the still-simmering prevalence of anti-German sentiment was apparently outweighed by the Nobel committee's attempt both to recognize the importance of his scientific work and to make a point in favor of forgiveness and reconciliation. Haber's was one of three Nobels awarded to Germans in the year following the war. Germany was being welcomed back to the world of international science.

The announcement of Haber's award sparked international outrage. Two French Nobel winners refused to accept their prizes in protest. An American Nobelist refused to attend any ceremony with Haber and physicist Max Planck (who won the prize in physics) until they disavowed their signatures on the 1914 intellectual's manifesto. The protests had no effect. In the view of the Nobel committee, Haber's prize was, if anything, overdue. He had been nominated many times, from 1912 on; he had received three nominations in 1919 alone. The ammonia synthesis was a great step forward for humanity, a potential boon for peace, a promise of future plenty. Its value could not be ignored for political reasons.

By the time Haber attended the Nobel ceremonies in the summer of 1920, he was well on his way to reassuming the role of international scientific leader. His beard was shaved, his demeanor kindly, his manner once again masterful. He delivered a lengthy Nobel speech in which he generously credited Robert Le Rossignol's contributions and spoke of the importance of Bosch's subsequent work (but did not mention Ostwald). He spoke at length about the importance of fixed nitrogen to agriculture (but ignored its role in warfare). He did not utter the words "gunpowder" or "explosives."

Then he returned to Germany, his work, and his increasingly unhappy marriage. He again threw himself into his professional life, busying himself with the many duties related to his position as institute director and adding new ones: German representative to international scientific organizations, professor of chemistry at the University of Berlin, head of the German Chemical Society. He worked late. He traveled often. "In my opinion," wrote his young wife, "you can't expect a twenty-eight-year-old woman . . . to have breakfast in a hustle and bustle at 8:30 a.m. and supper around 9 to 10 p.m. in the company of a man who's usually flat-out tired."

Haber also had a secret life at work. In the years after the war he maintained the high-level contacts he had made while working for the German military and continued his research into chemical warfare. He seemed to believe that Germany's foray into poison gas held promise for the future. He seemed to think that Germany would once again rise to greatness. The French, British, and American victors might be able to disband the German army, scuttle the German fleet, and shutter its munitions plants, but they could not stop chemistry.

Publicly, Haber's institute appeared to be engaged in researching ways to destroy stocks of war-related chemicals and use poison gas to kill insects and vermin. Privately, it was using the results from this work to enrich the understanding of chemical warfare. He kept a framed picture of the first gas attack at Ypres on the wall of his study. He stayed in touch with the German Ministry of Defense. He assisted in keeping at least some of the processes and plants developed for chemical warfare in working order, converting them to peacetime chemical production and sidestepping the Allied inspectors. At the same time he served as a secret adviser for nations who had been impressed with what the Germans had done and wanted to do it themselves. He became a sort of middleman for a shadowy business in illicit arms, fielding high-level inquiries about gas warfare and passing them on to a man he knew who had run a mustard-gas plant during the war. One of the interested parties was Spain, then fighting Riff rebels in North Africa. Contacts made between the Spanish military, Haber, and German army departments led to the shipment of needed chemicals from Germany to Spain and the construction of a mustard-gas plant in Spanish Morocco. Soviet Russia too wanted to get into the chemical weapons business. Haber helped put together people for a project to build a poison-gas plant on the Volga, the output from which, according to special instructions from Leon Trotsky, was to be shared between the German and Red armies. All of this was in direct violation of the Versailles accords. Haber did not much care. This was a new world, the world of the Weimar Republic, of a crippled and humiliated Germany, of shady deals and strange alliances.

Haber intended to do everything he could to revive and rebuild Germany. The quasi-military deals were part of that, but he soon had a bigger project going. At Versailles the victors had levied a crushing reparations burden of 132 *billion* gold marks. Haber knew that paying off the unbelievable debt would take away any chance Germany had of remaking itself. So he determined to discover a way to personally pay it off. Just as he had found a way to turn air into bread, he would now conjure all the gold Germany needed to pay its debts—from the sea.

EVERY SCIENTIST KNOWS that the world's oceans are a dilute solution of dissolved salts, trace minerals, and metals. The metals include gold, which, Haber read, was present in amounts around six milligrams per ton of seawater. It did not seem like much until you multiplied it by the number of tons of seawater in the oceans. The sea then began to look like a huge bank, with vast vaults holding millions of tons of gold, waiting for someone to come along and claim it.

A number of scientists already had thought the same thing but stopped short when they considered the difficulty of extracting small amounts of gold from enormous volumes of seawater. Not Haber. He read everything he could find about gold in the sea—the published research all pointed toward amounts between five and ten milligrams of gold per ton of seawater—and began to brainstorm a system for pulling it out. It would not be easy. Chemical means would be used, probably electro-chemical, some variation of the methods used to coat metals with metals, the way objects were plated with silver or gold. Haber needed to know more. He appointed a research group and set them up in a semisecret, restricted-access area of his institute. The others in the institute knew it only as Department M (for *Meerforschung,* "sea investigation").

Haber figured that he needed about fifty thousand tons of gold (an amount about ten times bigger than all the gold in Fort Knox) to pay off all the German war reparations with a bit left over. Department M scientists started testing systems to glean it from the sea. They quickly found that they needed extraordinarily accurate tools, ways to measure minute amounts of gold dissolved in water in order to verify the old studies and test for areas in the sea that might be richer than others, then a reliable, precise chemical system for collecting the metal. Measuring dissolved gold was extremely difficult when the amounts were as small as a few milligrams per ton of seawater. Tests could be thrown off by trace amounts of any metal coming into contact with the saltwater; something as small as a technician handling a gold coin could ruin measurements. They designed metal-free equipment, making their sinks from earthenware and making sure that nothing containing gold was present anywhere in the lab. Their first tests were made on saltwater they created in the lab and laced with known amounts of the precious metal. Slowly, over months, they developed a measurement system that accurately and reliably told them how much gold was in the water.

Then the researchers in Department M worked on ways to get it out. They tried an old method, bubbling in sulfur dioxide to combine with the gold and sink it to the bottom, but found that they could recover only about a third of the gold in the water. They tried lead acetate. They tried mercuric nitrate and ammonium sulfide. Finally, they settled on another ancient system called cupellation, an updating of methods known to the

Babylonians. Lead sulfide was used to precipitate the gold from the water, and the resulting mix of lead and gold (and silver) was purified by burning off the lead and then separating out the silver. The result was a small button of pure gold. Haber planned the next step: a voyage to sample the seas. He had to confirm the older studies with his new equipment, and there was the off chance that he might find an oceanic mother lode, an unusually rich gold region. It appeared that levels of gold varied from spot to spot in the sea. It was possible that under the right set of conditions, optimum water temperature and currents, dissolved gold might concentrate in certain areas, making it easier to collect. Searching would be expensive, however. He would need a complete laboratory, specially fitted, inside a ship. To avoid triggering competing teams from other countries, the work would have to be done in secret. He began quietly looking for financial backers, talking to a selected few of his many contacts in banks and industry.

IN THE SUMMER of 1923, the Hamburg-American liner *Hansa* departed Germany for New York City, carrying 932 passengers. Among the crew was a stocky, balding paymaster named Haber, who, together with a number of other crew members, disappeared for most of the cruise. They were in a secret room built into the liner's main deck, a specially designed laboratory outfitted with its own power, water, and gas, air compressors, crates of glassware, racks of chemicals, and unique equipment that would minimize contact with any metal. Within days of departure rumors began circulating around the ship: some mysterious men on board were testing new ways to propel ships, or harnessing forces to halt ships in midocean, or finding ways to create electricity from the water, or working on systems to prevent corrosion. Haber and his men started at least some of the rumors to divert attention from what they were really doing.

In the secret lab three chemists worked around the clock

analyzing water samples gathered as the ship steamed west. The gold they found was collected into small balls that Haber himself carefully analyzed and weighed, extrapolating just how much they could expect to harvest once the system became fully functional. He was disappointed at first. The early samples were puzzlingly meager, much smaller than he had expected. Perhaps the Atlantic waters here, near Europe, had lower concentrations of gold.

When the *Hansa* arrived in New York, Haber and his assistants, before debarking, reported on deck with the rest of the crew. Haber seemed to enjoy himself as he took the landing card handed to him by an immigration officer. The Nobelist was too well known, however, to remain out of sight for long. Reporters caught up with him and peppered him with questions. One asked Haber why there were so many paymasters on board the *Hansa*. "Because of the many zeroes," he quipped, referring to the inflation then rampant in Germany. He visited the General Electric laboratories and the Rockefeller Institute. He kept the press off the scent as much as he could. One paper published a story under the headline GERMAN SCIENTISTS SEE WAY TO DRIVE SHIPS BY USING MYSTERIOUS FORCE.

He returned to Germany worried about what he had found. There was much less gold in the water than he had expected, a hundred times, five hundred times less than the earlier published estimates. Either something was wrong with his testing system, or something was wrong with the old studies. Either way the news was not good. He reassured himself with the fact that this was a single cruise across a single section of a single ocean. He and Department M rechecked and refined their methods—which seemed to be working fine—and in the fall of 1923 he again took to the sea, this time with an improved shipboard laboratory, through southern waters, to Buenos Aires. A few months later he took a third, again to South America, sampling the warm currents around the equator.

But the results continued to be disappointing. Haber returned home and tried one more, world-spanning attempt, shipping out ten thousand carefully prepared, empty half-gallon bottles to scientists, sea captains, fishermen, naturalists, lighthouse keepers, and amateur volunteers, asking them to collect seawater and send their samples to Berlin. The Department M team analyzed them all. They all showed the same results: not enough gold. The amounts found everywhere were far short of what he needed to make his gold-recovery scheme feasible. After five years of trying, Haber had to conclude that the old published estimates were simply wrong. In 1927 he wrote, "I have given up looking for this dubious needle in a haystack." He did not even bother publishing most of the data. The reparations problem would be left for others to solve.

Chapter 16

CARL BOSCH UNDERSTOOD machines—he had an affinity for things made of metal—but he was less talented when it came to understanding people. This became an issue for him in his new role as head of the world's biggest chemical firm. He found himself in the early 1920s running a company blessed with the world's most profitable new technology, and cursed with some of the world's most challenging labor and financial problems. His first years of directorship were marked by crisis after crisis.

It started with the workers at Leuna, the great machine-city he had built during the war in central Germany. During the ferment that followed the kaiser's abdication, communist and anarchist organizers had made headway at Leuna, gathering groups of workers to talk about conditions at the plant, railing against the evils of for-profit business, and foretelling the end of capitalism and the rise of a workers' state. Like most industrialists, Bosch had a deep fear of labor activism in general and communism in particular. BASF had been built by men like Heinrich Brunck, who took care of their workers as fathers took care of their children, practicing a form of corporate paternalism that had worked in Germany for decades. Now this comfortable setup was threatened by workers who stood up to management, demanded better working conditions and pay, and threatened slowdowns and strikes if they were not treated fairly.

The workers at Leuna certainly had reason to complain. The giant plant, with its air-to-ammonia and ammonia-to-nitrate capabilities, had been built hastily during the war. Labor shortages had required bringing in thousands of workers from around Germany, many of whom knew little about construction. Housed in makeshift barracks, given an often cursory introduction to their sometimes dangerous equipment, and ordered to work long hours, thousands were injured during the construction. Forty-nine workers had died. Complaints were buried under a wave of wartime patriotism.

When the war ended, Bosch was faced with lingering unhappiness among his workers, and increased agitation by organizers. He was interested in a somewhat abstract way in labor theory, and he read books on socialism. He was among the first industrial leaders in Germany to institute an eight-hour workday and five-day workweek. According to one historian, he "devoted considerable effort to the improvement of relations with trade unions." But at heart he was a Brunck-style manager, more comfortable giving gifts than he was negotiating. He built new workers' housing to be as roomy and airy as possible, and he rewarded his most productive laborers with Christmas bonuses. His trouble started when he tried to sit down with labor leaders and deal with their problems. He had no stomach for long meetings and endless talk. The labor representatives seemed unable to appreciate the intricacy and balance of his great machines, and the need for humans to play their correct part in their functioning. Instead, workers went on for hours with their lists of complaints and demands. Long sessions would end with nothing being resolved. Bosch soon lost patience.

His general response was to institute a new American innovation called scientific management. BASF hired "efficiency experts" to observe workers as they did their jobs, men who took notes and critiqued every move just as they might assess the efficiency of cogs in a machine. They noted every time a man stopped to talk to his neighbor. They marked every bathroom break. They rated every action in relation to maximal production.

Not surprisingly, the workers hated it. In the spring of 1921 thousands of Leuna laborers, organized by tough communist activists, some armed with machine guns, took over Leuna and barricaded the entrances to Bosch's dream factory. There was a standoff for ten days. Then police using artillery started to fight their way in. A full-scale battle erupted. More than thirty laborers and one policeman died before the barricades were smashed and the factory returned to Bosch's control. Armed police swept through the workers' housing areas, arresting hundreds of strikers.

Bosch and his managers fired every worker at Leuna, without exception, and then rehired them one by one. Suspected troublemakers were culled out. To play it safe, the BASF management made it a policy not to rehire any laborer under the age of twenty-five. Bosch and his managers rolled back some earlier concessions and took away a few of the rights to organize that workers had recently won. New security measures were put in place. ID cards had to be shown by workers before entering the plant. In the early 1920s, with Germany adrift and the economy weak, people needed jobs. The plant was quickly back to full production.

Bosch felt both betrayed and perplexed. He wanted to be a good father for his company as Brunck had been. He felt himself to be a liberal, giving man, willing to listen to reason. He did not seem to realize that it was his own personality—uncomfortable around people, chronically devoted to efficiency, somewhat mechanical, often distant, coming across as brusque—that often got in the way of developing good labor relations. Times had changed and workers' expectations had changed. The old paternalism no longer worked. The labor revolt at Leuna only served to illustrate Bosch's shortcomings. He felt himself, perhaps not a savior, but at least a responsible businessman who had engineered a perfect machine at Leuna, a giant mechanism that was employing thousands upon thousands and helping keep Germany economically healthy. Why could his workers not be more grateful?

He was discovering that managing people was far more difficult than running a machine.

ON SEPTEMBER 21, 1921, the windows near Bosch started rattling and there was the sound of a distant explosion, a boom so deep he could feel it in his stomach. He was in Heidelberg, a dozen miles from Oppau, but he knew where the sound had come from, and immediately rushed to the ammonia factory. When he arrived he found, where part of the plant had been, a smoking crater three hundred feet across and sixty feet deep. Some buildings had disappeared, others were flattened, windows blown out, roofs ripped off. The nearby workers' housing area looked as if it had suffered an artillery attack. Families were on the street, sitting on piles of furnishings dragged out of their homes. Fires were being put out. Workers were pulling bodies from the rubble. His managers told him that many people were dead, perhaps hundreds, and many more injured.

Bosch was shaken, but he hid it. He talked with local officials. He conferred with his managers about the extent of the damage and its possible cause. He retired to his still-functional administrative office where, with pencil and paper, he began carefully and precisely working out the fastest, most efficient ways to get Oppau functioning again. Then he returned to Heidelberg, to the villa his company had built for him there, and spent the evening at his wood lathe, making spools for his mother's sewing machine.

The next day work crews began clearing the damage and making repairs. One end of the plant had been obliterated, but most of it was still operational. Friends and families of the dead raised a large wooden cross near the crater and decorated it with greenery. Workers and community members gathered, placing bouquets, mementoes, and messages around the base of the cross. Soon it was engulfed in flowers and fluttering bits of paper. Bosch's managers organized the repair work and started a formal assessment of the damage.

Four days after the explosion, Bosch gathered everyone at the plant for a memorial service. He was still off-balance, overwhelmed both by the scope of the disaster—it was now clear that hundreds of his workers had died—and by the feeling that somehow he had caused it. His perfect machine had gone awry and betrayed him and his workers. He was anxious too about the financial effects of the explosion, about caring for the families of the dead and injured, and about the chances of it happening again. Then he rose at the memorial event and looked into the faces of his workers.

> I stand before you today with a heavy heart. This is doubly hard for me as the builder of the Oppau Works, for it concerns my life's work to which I have been attached with every fiber of my heart, and whose growth I have witnessed from the beginning together with co-workers who have stood faithfully by my side, in joy and in sorrow, throughout those long years of development.

Speaking from the heart was extraordinarily difficult for Bosch. The crowd of employees, many of them in black, was silent as he spoke. He talked next of the great success they had together in "creating the nitrogen combinations that Germany so urgently needed for feeding its people," and how BASF had made this great stride because of being "first rate scientifically and technologically."

> Distant observers cannot even come close in imagining the entire scale and intensity of the scientific and technical

investigation and work that we had to accomplish over the course of almost thirteen years. Only those who actively participated can comprehend that. The most minute questions of detail had to be solved, all angles and secrets of natural forces had to be researched, before it was possible to master all the difficulties that were part of the task.

Perhaps he realized then that his people were there not to hear a technical treatise but a eulogy for the dead. He shifted gears to a more difficult subject, one that pained him personally: the betrayal of the machine.

And so the blow of fate hit us even harder—we who believed to have achieved this goal, I and all the hundreds of men who worked with me and had given their best—the blow which revealed to us in a shocking way that all our work and efforts had been mere futile human efforts after all, that nature had not let her last secrets be forced from her by levers and bolts, that in the end we always and again stand before the gate of uncertainty. It has not been professional error or oversight that led to the catastrophe. New, as yet inexplicable attributes of nature have made a mockery of our efforts. It was precisely the stuff meant to provide nourishment and life to millions of our fatherland, the stuff which we had produced and distributed for years, which suddenly proved itself to be a cruel enemy for reasons we do not know. It has put our work to ashes.

This was the core of a paradox that was tearing Bosch apart. He had spent his career making things work, and now, inexplicably, there was destruction. The speech was a terrific strain for him, more so as he moved from the technological to the human side.

But what is all of that in comparison to the victims that this catastrophe has claimed. Here we stand totally helpless and powerless, and all the support we can give in order to comfort the grieving families and the injured is nothing in comparison to the loss. Only empathy and gratitude for what the dead have meant to us remain. As the representative of the directors and the Board I express my deeply felt grief.

Bosch tried to rally his workers the only ways he knew how, with appeals to the fatherland and their own economic survival.

From time immemorial mankind's battle with natural forces has claimed innumerable victims, mostly less noticeably, because they didn't quite reach our awareness. But here, in the face of an enormous catastrophe, this battle reveals itself in its entire, horrific tragic. For this battle is not voluntary, it must be fought, and even today, even before these open graves this inexorable "we must" already forces us back onto the path of fulfilling our duty. If there is anything that may comfort us in our misery, then it is the awareness that the tough tasks that continue to await us will serve the well-being of our fatherland, whose battle for its existence today is harder than ever as the consequences of the war become more and more evident. And one of the most important factors and conditions for the possibility to survive at all are our nitrogen works. . . . If I didn't know that despite the shock over this horrible accident the trust of our co-workers remains, I would have to despair over the new challenges that are facing us. For the dead, though, who have left us, who have descended into the dark realm of shadows, I have placed, in grateful memory of their faithful collaboration and fulfillment of duty, with a deeply stirred heart, a wreath on their graves.

The short speech, the most personal statement of Bosch's on record, is a testament to the opposing forces that drove and hounded him. On one side was his scientific pride, his utter belief in technological solutions to human problems. On the other was his human emotion, the shock he felt when his technology failed. There is the sense that his company was somehow responsible for the grieving families before him. At the same time, he refused to accept full responsibility and pushed to get his workers back to their jobs. On one hand were the triumphs of science; on the other, the secrets of nature that science cannot know. On the surface, he was quiet and as efficient as he could be; but on his way home after the memorial service, Bosch collapsed.

There is little good information on exactly what happened to him, whether it was more physical or mental, whether it was exhaustion or a nervous breakdown. All that is known is that he disappeared for several months, staying in seclusion through the winter of 1921 and into the following spring. It is also unclear whether, during those months, he began drinking heavily. What is known is that in the years that followed, his drinking became more frequent and the effects more noticeable.

While Bosch recovered, his lieutenants at BASF tried to figure out what triggered the explosion. BASF's executives wanted to know if worker agitation, perhaps sabotage, had been to blame. The workers wanted to know if they were working inside a huge bomb that might go off again. French and British observers wanted to know if the explosion signaled that Oppau had been secretly producing military explosives for Germany. Fritz Haber, who felt the shock of the explosion—"equivalent to an earthquake," he said—in Frankfurt, opined that it might be due to "something new in explosive forces."

Plant examiners determined that the blast had started in a storage silo, essentially a huge pile of fertilizer. One minor problem they had faced from the start of Oppau was that granular fertilizer tended to cake in storage, to absorb water from the air and solidify into a rocklike mass. Workers at Oppau had long used explosive charges, small ones, to break up the huge piles for shipping. It was thought to be absolutely safe: Oppau's primary fertilizer product was ammonium sulfate, and tests showed that ammonium sulfate would not explode. The small blasts had been used for years without a problem.

What was safe for ammonium sulfate, however, was not safe for Chilean saltpeter (sodium nitrate), the white salt that the plant had started producing for munitions during the war. Many farmers still preferred the reliable old Chilean product to BASF's ammonium sulfate. Rather than shutting down the production of white salt completely after the war, Bosch had continued making it, mixing it into his ammonium sulfate fertilizer. Tests by BASF chemists had indicated that the nitrate-sulfate mixture too was safe to store and ship as long as the proportion of nitrate was not too high.

They were wrong. Maybe the conditions deep in the silo were different than those in the lab. Maybe there was an unmixed pocket of pure nitrate somewhere in the pile. Whatever the particulars, one of the little blasts used to break up the pile triggered a huge explosion, like setting off a blasting cap in a bundle of dynamite. The entire silo went up, some forty-five hundred metric tons of fertilizer. The force of the blast was hard to gauge; one estimate put it around that of a small atomic bomb. While Bosch recovered, the company totaled up the damages: 561 workers dead, 1,700 injured, around 7,000 homeless, and half a billion marks in repairs to the plant.

Bosch's ultraefficient lieutenant, Carl Krauch, shouldered the job of getting Oppau back into full production. He achieved his goal in three months. Less than a third of the cost was covered by insurance. The company steadfastly refused to accept legal responsibility but agreed to pay onetime sums to the families of full-time workers killed and injured in the blast, adding small pensions for widows. The families of the many casualties who were not full-time employees—mostly contract workers for construction or shipping firms—got payments worth (given the value of the mark at the time) less than fifty dollars.

"As these agreements could not satisfy everyone," a history of the company noted, "they long remained a source of discontent and mistrust toward the firm." After the rubble was cleared, communists won control of the Oppau workers factory council in 1922—but not for long. Following the firing of three BASF labor leaders who dared to attend a communist-organized national labor meeting, Oppau's workers went on strike. When the strike was broken several weeks later, the company's directors purged thirteen hundred strikers and then, as at Leuna, tightened the rules governing those who were left. Oppau remained through the early 1920s the scene of a string of protests, dismissals, strikes, and occasional killings.

BOSCH RETURNED TO work in the summer of 1922. He seemed the same on the surface, but something underneath had changed. He had always been uncomfortable around people, but at least had been able to talk easily with engineers and machinists, to go bowling on the weekends and tell jokes. He had occasionally shown a playful side, a fondness for odd practical jokes, like the time he put a live fish in a bathtub at a party. After the Oppau explosion, the jokes ceased. He became serious, quiet, and withdrawn. He cut back on meetings and whittled down the size of his top advisory councils until he could make the biggest decisions by meeting with an inner cabinet of just a dozen or so men. He gave his lieutenants ever-wider authority to decide matters independently. He began spending more time in his Heidelberg home. He spoke only when necessary and was, an observer noted, "relentless in insisting that his associates stick to the point." He had, by some accounts, started drinking more heavily in private. He was forty-eight years old.

CHAPTER 17

IN MID-MAY 1923 Carl Bosch learned that the French again were marching toward Oppau, looking for reparations, planning to occupy his factories on the Rhine, and intending to arrest him. He had one day to prepare. He immediately shut down operations at both Oppau and the original dye factory at Ludwigshafen. As before, the French would stop at the river, which meant that Leuna was safe. But the two plants on the French side of the Rhine were not. Within hours his people were pulling apart the huge ammonia ovens at Oppau and ferrying pieces across the river, getting as much of the high-pressure machinery as they could onto trains to Leuna. They began emptying warehouses full of dyes and finished chemicals, carting the goods across the Rhine bridges. The firm's management and board members, including Bosch, fled to Heidelberg. Bosch left three deputy directors behind to deal with the occupation.

The Germans had stopped paying reparations, the Weimar government said, as a matter of survival. The nation had been hit during the early 1920s with one fiscal crisis after another, and as part of a short-term solution had made the mistake of printing a great deal of money. The flood of paper marks kicked off a period of hyperinflation in the early 1920s during which the value of the mark plummeted so far, so fast that during the worst of it the mark essentially lost all value. Workers were paid with bundles of currency thrown off the backs of trucks, and ran to spend their wages on food before prices went up again. A pound of bread could cost eight hundred million marks. A pound of butter might cost a billion. The poor could hardly afford to eat; the middle class watched its savings and pensions dissolve. At one point BASF resorted to printing its own money for its workers, "aniline dollars," backed by the company's bonds and foreign deposits.

In the face of this newest crisis, Berlin announced late in 1922 that it had no choice but to suspend reparations until the inflation was under control. Some of the recipient nations gave Germany time to grapple with its problems; France did not. French troops were sent in to occupy Germany's industrial heartland, the Ruhr, and eventually extended their occupation to the area of Bosch's factories where they took everything they could, from payroll transports, wood supplies, and coal shipments to finished dyes and chemicals. They emptied BASF's fertilizer storage silos at Oppau. Then they began taking iron, steel, glass, and cement, shipping it back to France to rebuild war-damaged housing.

Bosch—just as he had during the first occupation—powered down his factories and refused French orders to restart them. The French expelled uncooperative BASF labor leaders (for the moment, the occupation brought together German labor and management in a common cause) and held the remaining BASF directors as hostages. Then they charged Bosch and BASF board of directors with refusing to cooperate with their lawful authority and for refusing to supply the electric power they needed for removing nitrates from Oppau. The BASF executives were tried in absentia in a French court. All of them were found guilty, fined, and sentenced to prison terms. Bosch got eight years. He, along with the rest of the directors, refused to accept the validity of the sentence and stayed just out of reach, across the river in Heidelberg, from where he used the media to needle the occupiers. "The French may be able to make bricks," he told a reporter during the occupation, "but never dyestuffs."

The standoff continued through the summer and into the fall of 1923. BASF was hemorrhaging money because of lost production. Then the French, having long since taken everything of value, realized that the occupation was costing more money than it was worth. There were other forces at play too: The German central bank brought inflation under control by issuing a new currency backed by real estate rather than gold, and the United States was stepping in to broker a new reparations plan. A newly elected German chancellor promised to resume reparations payments.

Toward the end of the year, the French went home, and Bosch restarted his plants.

THE PERIOD OF hyperinflation was not all bad for BASF. The company had long been worried about repaying the millions of marks it had borrowed from the government during the war to build Leuna. When the inflation hit, BASF was able to pay off its enormous debt in greatly devalued marks, essentially paying back pennies on the dollar. Because so many of BASF's products were sold in other nations, the company profited from a stream of desirable foreign currency. The weak mark even allowed the dye companies to stave off growing foreign competition, because it allowed them to keep their prices relatively low. They did fine in general during the inflation, with enough resources to keep workers satisfied with raises and emergency aid, to feed them, and keep them away from the communist organizers. They came out of it in fine shape.

By 1924, with the French gone, inflation under control, and farmers buying fertilizer again, BASF entered a period of record profits. Two-thirds of the income came from Bosch's ammonia plants. There was money to pump into research and development, and money to invest in improving Oppau and Leuna. Every refinement made in the plants led to smoother production, greater automation, and higher profits. It happened just in time. Bosch had the best ammonia plants in the world, but they were no longer the only ones. The French now had a Haber-Bosch plant, the fruits of Bosch's secret negotiations at Versailles, which they were learning how to run without German help. It did not take long for the technology to start working its way to other nations. The British, who had been negotiating with BASF to license the technology, found it more expedient to deal with two Alsatian engineers who had appeared in London in 1920 and approached the London chemical firm of Brunner Mond, claiming that they knew all about Haber-Bosch—and were willing to sell what they knew. "The legality of their offer was, to say the least of it, doubtful," wrote a historian of British chemistry. The British negotiated nonetheless. The two engineers, called "K and A" in Brunner Mond documents, settled into a pleasant life in Paris while the deal making was done. The British chemical company's executives considered their reputations, the legality of what was being offered, and the reliability of K and A. They looked at the drawings and plans K and A had smuggled out of Germany. They thought about the prospect of interminable negotiations with BASF. They thought about the high licensing fees. Then they decided to buy the stolen material.

They got a deal. K and A provided what they had promised, providing the British with schematics from Oppau and Leuna, along with detailed estimates of production costs and comparisons to other techniques. "The information we have obtained from K and A since the contract was signed seems to us of the very highest value and importance," exulted one British chemist. For this the two Alsatians were paid 500,000 francs down (about 8,500 pounds sterling), plus 750,000 more to be received when

540.92 H

the British got a plant up and running. Within five months the British were producing Haber-Bosch ammonia in a pilot plant. By Christmas 1923 they had two plants running. Workers celebrated with a fancy dress dance where they sang a holiday carol with the line, "and ammonia was made on Christmas Day." They did all of this without paying BASF a shilling in licensing fees.

The Haber-Bosch secrets continued to spread. There was now a lucrative market for German ammonia engineers. The United States had spent millions trying and failing to build a Haber-Bosch plant during the Great War. Afterward, DuPont, taking a page from the British, lured away its own high-ranking German technicians. With their help the United States too was soon making Haber-Bosch ammonia. BASF's great gamble was about to be lost. Instead of making licensing income, the company was facing increased competition. The more Haber-Bosch plants that came online around the world, the more the price of ammonia (and the fertilizers made from it) dropped.

Bosch had seen the day coming ever since Versailles. He might have tended toward depression and withdrawal, but he was still a step ahead of his competition. While foreign firms were figuring out ammonia, his own chemists were looking for new ways to use high-pressure chemistry to make new products. Their first breakthrough was a high-pressure method for making methanol (wood alcohol), a valuable bulk chemical required for a variety of chemical processes. Bosch's right-hand man, Carl Krauch, speedily put methanol into production at Leuna, giving the giant plant a new source of income and a new reason to be expanded. Methanol proved that high-pressure chemistry could be used for more than just ammonia. But it was not enough for Bosch.

In late 1923 Bosch took a long trip to the United States. Chemists in the two nations were very interested in each other. U.S. chemists saw Germany as a vital step in their education (they flocked to German universities both before and after the war to study with German master chemists), and German chemists saw the United States as their only real competition for industrial dominance. Bosch was thinking about fusing Germany's dye and chemical firms into a single big organization and wanted to see firsthand how giant U.S. businesses like Standard Oil operated.

He came home with new insights about the future. Everywhere he looked in American cities, he saw streets jammed with automobiles. Everybody wanted one. The United States was car crazy, in great part because Henry Ford had figured out a way to mass-produce them cheaply enough to bring them within the reach of millions of buyers. Automobiles were becoming a huge business, two million cars produced and sold in the United States in 1920, rising to an estimated three and a half million in 1923. Surging sales were accompanied by the rise of entire new industries: tire manufacturing, road building, auto insurance, lacquer making for finishes, repair shops for maintenance, batteries, lamps, upholstery, and the list went on. Much of the industry involved chemistry in some form, from making better paints to making better rubber for tires. But the richest area to explore, Bosch realized, did not involve automobile parts. The real money lay in what made them go.

The future belonged to gasoline. Gasoline, and the natural oil it was made from, was where the really big money would be. His talks with Standard Oil convinced him. He liked the people at Standard, who were turning what had been primarily a refining operation into an integrated oil company, from drilling wells to filling cars. They thought big, as Bosch thought big. But they were worried, too. In the early 1920s, virtually every expert was predicting that the world was about to run out of oil. The best estimates of existing oil supplies indicated that the major fields were going to be tapped out by the 1930s. Despite a worldwide search for more, no big new fields were being found. The world was about to face its first oil crisis.

Then a solution to everyone's problems came to Bosch: gasoline. He would make synthetic gasoline. The world would be hungry for it, it was basic, everyone needed it, it would sell in immense quantities, and, best of all, he believed it could be made by the millions of barrels using high-pressure chemistry. He was familiar with the research of Friedrich Bergius, an entrepreneurial German chemist who had tinkered for years with the idea of making new products out of coal. One system he developed involved grinding up coal, mixing it with crude oil, heating it, and putting it under pressure with hydrogen. The resulting "hydrogenated" coal slurry was altered chemically in ways that improved the yield of a variety of products. One of them was high-quality gasoline. In essence, the technique turned coal (which Germany had in abundance) into crude oil (Germany had no oil wells). Bergius had tried for a decade to turn his laboratory discovery into a factory-sized operation, but problem after problem cropped up, bogging him down until his backers lost interest. By the early 1920s, the inventor was ready to give up.

This was Bosch's chance. A grand vision was forming: If BASF worked with Bergius's process and perfected a high-pressure way to make gasoline out of coal, it would not only make big money for BASF, but would nurture the automobile industry in Europe, increase the number of cars, and thus boost demand for more gas. If the world ran out of oil, BASF would corner the fast-growing market. It could, perhaps, make deals with automobile manufacturers that would be of mutual benefit, giving the company an entrée into the car business, with all the money it promised. At the same time, synthetic gasoline would help make Germany energy self-sufficient in the new automobile age, just as ammonia had made it self-sufficient for fertilizer and explosives. Using the national self-sufficiency argument, he could likely get government support to cover at least some of the costs of development—including the expansion of Leuna.

Leuna was always at the forefront of Bosch's mind. If he could change it from an ammonia plant into the world's first high-pressure factory for making synthetic gasoline, he could not only keep his dream factory alive but could grow it, transform it, and make it the world center for a new technology.

This was going to be a major gamble, bigger in many ways than ammonia. The Bergius system relied on many of the same continuous-flow, high-pressure approaches as ammonia production, but promised to be far more difficult to carry out. The good news was that the temperatures required were somewhat lower than those for the ammonia process. The bad news was that everything else was going to be harder.

For one thing, it would eat mountains of coal, which meant that BASF would have to buy mines, operate them, and build new rail lines to the factory. Then the company would have to create and construct new systems for storing and grinding and mixing and moving the coal once it arrived. All that infrastructure would be expensive, and much of it would have to be designed almost from scratch. Then there was the catalyst. Bergius had failed, in part, because he had never found a good catalyst. That search too was likely to be costly. Most troubling was the fact that everyone at BASF was now accustomed to working with gases, and the Bergius process started with a very different sort of material, a messy, thick slurry, which would have to be fed somehow into a very different sort of reaction chamber. Bergius had gone a good way down that road but had stopped far short of the efficiency BASF would need. All the input and oven designs would have to be brainstormed. There was no telling what sorts of problems would crop up.

But the more he studied the process, the more Bosch felt that he and his teams could make it work. They had made ammonia work by solving problem after problem. They had turned Haber's tabletop machine into the giant plant at Leuna. Thanks to Bosch's leadership, BASF now had the world's best high-pressure laboratories with some of the world's best scientists running them. If he focused them on the Bergius process, they would meet the challenge. They would supply the world with synthetic gasoline.

The only thing he needed was money. The research alone would be staggeringly expensive; the conversion of Leuna would likely put the whole idea well beyond BASF's capabilities. The company alone could not afford it. The government might help, but the German government was still weak, still crippled by reparations. To fund his vision, Bosch would need more. He would need international deals with some of the world's biggest companies. More than that, he would need the backing of all of Germany's biggest dye firms. BASF was a major company, but it was small compared with firms like Standard Oil. If he wanted to deal with the big boys on an equal footing, he needed more financial power behind him. He went back to the idea of banding the dye firms into a superbusiness big enough to secure funding for his synthetic gasoline project—and one big enough to dominate the global chemical market.

BOSCH WAS NOT the only one thinking about making a superbusiness. Talks between the independent German dye and chemical firms had been going on, in one form or another, for years. Various temporary alliances had been made at various times, most recently during the war. Carl Duisberg, the head of Bayer, another of Germany's top-three companies (along with Hoechst and BASF), was strongly in favor of merger. But the old dye companies were proud of their histories and their product lines, protective of their patents, and had always resisted taking the final step.

In the 1920s, however, the ground began to shift. Part of it was technological: BASF's success with high pressure had shown that big machinery and big investments were going to be

needed in the postwar world. Corporations needed to join together if they were going to build vast new plants using expensive new technology. In Britain four firms banded together after
the war to form the Imperial Chemical Industries. In the United
States five firms joined to make Allied Chemical. If the Germans were to compete, they would have to do the same. When
the German government began to pass laws in the early 1920s
that benefited the formation of larger business entities, it became clear that the time was right to merge.

Duisberg supplied the site for a summit meeting between all
of the leaders of the German chemical industry. The Council of
the Gods, as they called it, took place in November 1924 at his
private villa north of Cologne. Everyone who mattered was
there, including Bosch. The two-day summit, clouded in cigar
smoke, fueled with fine food and the best liquor, and marked by
arguments over merger details, gradually evolved into a competition between Duisberg and Bosch. The Bayer head champi-
oned a detailed total merger, while BASF's chief promoted a
looser cartel structure. The Duisberg backers gathered in the
billiard room, the Bosch faction in the villa's bar, with a media-
tor rushing proposals and counterproposals between them.

In the end, most of the attendees sided with Bosch. Many
details would have to be ironed out before the papers were
signed, but a general plan was set: Germany's biggest dye and
chemical firms would create an "interest community" to coordi-
nate research and sales efforts. They would merge a number of
administrative functions. They would present a united front to
the world—while still marketing and selling products under
their old individual names. It became official in the fall of 1925
under the name of the Interessengemeinschaft Farbenindustrie
Aktiengesellschaft (literally, the Interest Community of the Dye
Industry, Inc.), an unwieldy name that the public quickly short-
ened to IG Farben, or, simply, Farben. Farben was, at the moment of its birth, the largest business in Europe, the largest

chemical company in the world, and the third-largest business organization of any sort, measured by the number of employees, on the globe (bested by only U.S. Steel and General Motors). Bosch was named its director.

It was the structure he needed to back his synthetic gasoline project. Just as Farben was being launched, Bosch bought the Bergius patents and started his teams to work.

Then he settled into a lifestyle commensurate with his new position. In 1921 an enormous villa had been built for him by BASF on a hillside in Heidelberg, with a sweeping view of the Neckar River valley, a short walk from the ruins of the old Heidelberg Castle. The massive home was a showpiece designed to highlight the aristocratic nature of industrial power, with elegant public rooms, a separate chauffeur's house, and immaculate lawns and gardens. Bosch stamped it with his personality by installing a fully equipped mechanical workshop, a study lined with cabinets for his mineral and other collections, a complete chemistry laboratory, a physics lab, a darkroom (photography was a favorite new hobby), and an astronomical observatory in a separate domed building, equipped with one of the largest private telescopes in Germany. Bosch hired an assistant to oversee his growing collections of plants, minerals, and animals. He hired another to tend his laboratories. Then he hired another to tend his telescope.

Despite all the distractions, he still suffered from melancholy. He still did not much like the people side of his business, and now that he was head of one of the world's biggest companies he was obliged more than ever to travel to other plants, hold business meetings, and host social events at his home. He was never very good at it. One guest remembered arriving for a dinner party and chatting for a very long time without a glimpse of his host. Finally, much later, dinner was announced. The guests sat, waiting for Bosch, while his wife Else kept the conversation going. They waited ten minutes, then twenty, then thirty, with Else acting as if it were all routine, before Bosch finally appeared. He apologized by explaining that he had been unavoidably delayed: On his way to dinner he had passed by a grandfather clock in the hall that was out of adjustment. He had started dismantling and cleaning the clock and had, well, lost track of time.

Instead of business, he preferred being by himself or with an assistant, organizing his collections and tinkering in the lab, or in the company of scientific men with whom he could discuss chemistry, astronomy, or zoology. When talking about business, he could be abrupt. When talking about basic science, he lit up. Bosch kept a large private library and read the latest scientific literature. He knew enough about cutting-edge chemistry to form a friendship with Walther Nernst, and he was current enough in physics—not even his field—to be able to chat knowledgeably with Albert Einstein.

He grew to love astronomy and became a night owl, spending clear winter evenings in his unheated dome, wrapped in blankets, wearing fur boots, peering through his telescope. Once he rushed to the house and woke his wife, ushering her outside on a particularly frosty night so she could see, up close, a strikingly attractive crescent of Venus. "If not constructing a new apparatus in the workshop of his home in Heidelberg, or investigating astrophysical problems in his private observatory, he most likely sits over his beetles or plants," Alwin Mittasch wrote about his often-absent boss.

Bosch too was an exquisitely balanced mechanism. He needed his private time if he was to function publicly at Farben. He needed order and serenity to offset the stress of crises and decision making. He needed time to play with his collections to vent the pressure. He maintained that balance through the formation of Farben and the rest of the 1920s. A few more years would pass before it all fell apart.

CHAPTER 18

FRITZ HABER, MEANWHILE, was focusing on building his research group. After the failure of his gold-from-seawater scheme, his institute became his life. Good at spotting scientific talent, he hired productive researchers and gave them a relatively free hand to develop programs in some of chemistry's most exciting and productive fields. During the 1920s their work helped push chemistry to the borders of biology and physics, yielding important findings about the nature of colloids, chain reactions, charged particles, gases, and flames and explosions, among other things. Haber's institute became an international mecca for physical chemists. He started a regular series of colloquia in which visiting experts would give talks followed by free-ranging discussions in which institute members and guests argued points, followed up insights, and, as one attendee put it, talked about everything "from the helium atom to the flea." Haber set the tone with his quizzical expression, erect posture, big cigar, and provocative manner, asking pointed questions, making jokes, and poking holes in reasoning. At home he could be a demanding tyrant; in military circles he could be a ruthless warrior. But in his institute he was a genial father figure. Voices rose in Haber's colloquia, people laughed, and thinking changed. Students in particular came away enthused by Haber's iconoclastic approach. He exemplified what they came to call the spirit of Dahlem—in honor of the Berlin suburb where Haber's institute was located—a combination of rigorous study and open minds. Dahlem, with Haber's institute and the others started before the war, was turning into Germany's Oxford, as one historian noted, remote, a little eccentric, and devoted to excellence, a place that one visitor during that time recalled as "the very empyrean of science." Haber was a central figure at Dahlem, his wartime infamy eclipsed by his new role as an elder statesman. Inside the institute he was like a brilliant favorite uncle, bald and genial, keeping up a stream of jokes and stories, but getting to the heart of scientific issues quickly and unerringly, attacking scientific problems "like a bull out of the gate," as a British scientist noted. Outside the lab he was an exemplar of German excellence, his bearing erect and collar starched, a respected proponent of scientific advancement and international communication. He was, one young colleague wrote, "larger than life in every sense." The physicist Lise Meitner had a slightly different impression. Haber, she said, wanted to be "both your best friend and God at the same time."

All his energy was spent at work. By the time he got home late in the evening his young wife Charlotte saw a different man, serious, exhausted, worried about illness, and unhappy with his life. She told him how she felt after returning from a trip she took (by herself) in the early 1920s: "There are dark shadows in this house," she wrote Haber. "There's no room for harmless jokes and fun." By the early 1920s they had a second child, a son, which only seemed to increase Charlotte's sense of alienation. In 1924 the two of them tried to patch their marriage by taking a half-year cruise around the world, just the two of them. Charlotte loved the time away, but their reconciliation did not last much beyond their return to Berlin. "I either want to direct things," Haber admitted around that time, "or let them go." He was writing about his research, but he might as well have been talking about his marriage.

By 1927 it was over. "I feared marriage with you because our natures are completely different," Haber wrote Charlotte at the end. "Your friends are not my friends; your inclinations are not mine. Even when we are together, we live for our individual selves, and your attempts to change are as futile as mine. Let us call ten years enough. I can't do it anymore." They were divorced in December, three days before Haber's fifty-ninth birthday. He viewed it as a great personal failure. The end of his second marriage "gnawed at me and humiliated me in my own eyes," Haber wrote Einstein, leaving him with "long days in which I am wholly filled with a sense of superfluity and mediocrity." He fell into depression.

His response, as usual, was to throw himself into his work. He expanded his already heavy schedule of international meetings and professional obligations and worked to his physical and psychological limits. He suffered from insomnia. His health continued to decline. His doctors told him to take time off, to ease the stress; his heart, they warned, was giving out. He started taking nitroglycerin to ease his chest pains. He would recover a bit at a health spa or on vacation, then come back to work and, after a few months, fall ill again.

As his health declined, Haber began to think more about being Jewish. He had always tried to set matters of religion and race aside and focus on assimilation. But he was still a Jew. Conversion for Haber had been a convenience, not a commitment. His social circle, his best friends, and his wives were Jewish. A Jewish financier, Leopold Koppel, had financed his institute. Between a third and a quarter of the people who worked there were Jewish, a far higher proportion than the general population. He did not talk about it often, but he did in 1924 when faculty at the University of Munich objected to the appointment of a Jew to the faculty. Outraged at the rejection of a clearly superior candidate, Haber's longtime friend Richard Willstätter resigned his professional post at Munich in protest. Haber wrote immediately to support his friend's decision, calling it "a blow that will echo throughout the world." He also made mention of the resurgence of increasingly open discrimination against Jews and linked it with the name of a certain right-wing trouble-maker. "I understand that in [this matter]," Haber wrote, "you have encountered more Hitlerism than you can bear."

But Haber had other things to worry about in the 1920s: his institute, his health, and his money. He kept up a whirlwind pace of organizing and traveling, watching the quality of work coming out of his laboratories, and keeping money flowing to the Institute. His heart continued to weaken, and the necessary spells of rest grew longer. After his divorce, his financial situation became difficult; he had promised to support Charlotte and the children, and she was accustomed to a high style of life. He had taken a lump-sum settlement from BASF some years before, ending the income from ammonia. Then he lost some major investments he made in South America. He still made a very good salary, still smoked the best cigars, ate at the best restaurants, and stayed at the best hotels. But he could not stop worrying about money.

LEUNA, BOSCH'S MECHANICAL masterpiece, looked from a distance like a small city built around a line of thirteen very tall smokestacks. It was actually a single integrated machine, a complicated assembly line two miles long and more than a mile wide. After the worker revolt of 1921 it was run like a security facility, with its own police force, walls around the perimeter, and guards at the entrance gates. Each of the thirty thousand or so workers who went in and out every day showed identification cards. Every day a few of them were pulled aside for random searches; they were patted down and their pockets emptied.

To find out what it was really like, an enterprising German journalist went undercover as a Leuna laborer for a number of days in the 1920s. "A great airy steel construction, branching and interweaving in a huge labyrinth of metal," he wrote, an "inextricable tangle of iron, stone, wood and cement" that reminded him of a "giant iron worm." He felt overwhelmed among the huge steam pipes, retorts, conduits, siphons, compressors, ovens, elevators, purifiers, filter presses, absorption chambers, cooling towers, gasometers, mills, and electric lines, rows of generators "resembling a rank of giants on parade," and fields of slag "like whale backs in the steamy distance." One moment he would be working a job amid roaring, whistling, hissing clouds of steam and occasional jets of colored flame. The next he would be assigned to some great industrial hall filled with pipes, dripping towers, and pumps, the whole of it "as still as a churchyard."

The way he described it, Leuna was a hell. To increase productivity, workers were assigned piecework, paid not by the hour but by the task. To make more money they rushed through job after job, working themselves into exhaustion, taking sick days every few weeks to recover. The report filed by the journalist sounded like scenes from Fritz Lang's *Metropolis* (which was released in the mid-1920s, during Leuna's heyday).

Carl Bosch, of course, saw it differently. From his villa in Heidelberg, Leuna was more a triumph of human ingenuity than an industrial inferno. Yes, his laborers worked hard, but they were showing what humans were capable of. They were in fact among the most efficient workers in the world, and they were adequately housed and paid through the worst of times. Leuna was his great achievement. His foremost goal was to keep it running and expanding as long as he could.

The problem was that the giant factory was far more than a technological wonder. It continued to be, in the view of some Farben board members, a sinkhole for money and, through the 1920s, a financial disaster in the making. Bosch had pushed for the formation of Farben in part to create an organization big enough to fund Leuna's expansion, the last phase of which had been planned back in the final months of the Great War. He wanted the plant completed no matter what the cost. As the 1920s went on, Leuna seemed to become an obsession for Bosch. The larger it got, the more power it seemed to hold over him.

The man and the plant were inseparable. It was not only because Leuna was the greatest expression of Bosch's mechanical genius. He believed that it was also the key to the future. There was only one way for BASF to stay on top: ceaseless invention. Dyes were on their way out, fixed nitrogen was on its way out, he needed the next big thing, and that was going to be synthetic gasoline. Leuna was the perfect place to make it, centrally located, near coal mines, good rail connection, plenty of water, and most important of all, a good infrastructure for doing high-pressure work. The company had plowed nitrogen income into high-power research laboratories, in-house development centers that were starting to work on synthetic gasoline. When investors complained that their dividends were disappearing into the labs, Bosch responded, "I consider it a far higher moral obligation to provide a secure livelihood to the . . . men and women currently employed by our firms rather than respond to a fluctuating economy or fluctuating revenue and simply pay out a dividend when possible." The payoff was going to come from Leuna's successful creation of synthetic gasoline.

He had no doubt it would succeed. "The future of the chemical industry is unlimited," he told a journalist. "There is no natural product that cannot, with perseverance, be manufactured." The future belonged to synthetics. The future belonged to Farben.

In 1926, at one of the first meetings of the newly formed megacompany's management, Bosch outlined Farben's future. Success, he said, would come not only from great technological breakthroughs like synthetic fuels, but would also depend on Farben's breaking out of Germany and moving aggressively into the world, stepping up its scale of operations in other nations, especially in the United States. He did not need to add that much of the world still despised Germany—his directors knew that—making it necessary to position Farben as more than a German company. To protect itself from political and economic instability at home and to thrive abroad, Farben, Bosch said, needed to aggressively create foreign subsidiaries and forge high-level alliances with non-German companies.

The key to doing that, again, was synthetic fuels. This was his vision for the company: Let each of the members make their products and profit from them, but let each give enough to the central enterprise to fund Leuna's, and the company's, leap into synthetic fuels. There was discussion; there was some internal grumbling. But in the end, they followed Bosch's lead. Money started to flow to Leuna, and the great conversion to synthetic gasoline began.

Later in 1926 Bosch returned to the United States with an entourage of Farben executives, ready to introduce Farben to the world and ready to make deals. His first step was to reach out to the other big players in chemistry. He had already been in talks with Sir Alfred Mond of Brunner Mond and with others involved in the emerging Imperial Chemicals group in Britain, and he had opened communication as well with representatives of Allied Chemical in the United States. He told them that Farben was quickly improving the Bergius process and that the company would soon be making synthetic gas in bulk. One of his top young chemists had already come up with a much better catalyst. Other technical problems having to do with the preparation of the coal and its movement through the reaction chamber were being solved, one by one, by his research teams.

Bosch was using a different strategy than Heinrich Brunck had used with ammonia. Back before the war, their idea was to keep the technology as secret as possible so that competitors would have to license it. That strategy had failed. Now Bosch, on his 1926 travels, seemed eager to share the technology even before it was complete, dangling it before other companies as it was still being perfected, seeking lucrative deals.

He needed the money. It was true that his research teams were making progress, but they were encountering more problems than they had foreseen, and the costs were mounting. Bosch believed that making synthetic gasoline and other fuels from coal was going to be the most lucrative new technology of the century. So he began offering deals: If other giant corporations wanted in on the deal of the century, they could form partnerships by paying the admissions price in cash or an exchange of stock. Bosch intended to use the income to pay for the final development and scale-up of his synfuels plant at Leuna.

It was tempting. The British became deeply interested in synthetic fuels as soon as they heard about it. Brunner Mond had spent years failing to duplicate the ammonia process before the company bought a set of stolen plans; the firm's chemists were properly respectful of the Germans' technological abilities, and appeared eager this time to make an arrangement. An enthusiastic Sir Alfred Mond sailed to the United States in the fall of 1926, presumably to get Allied Chemical to join with him in buying into a deal for sharing Farben's synfuels technology. The deal between the world's top three chemical firms was supposed to be ready for Bosch to sign when he arrived a few weeks later.

But it never happened. At the last moment Allied backed out, deciding instead to cozy up with Farben competitor DuPont; those two in turn developed a close relationship with General Motors. Instead of signing with Allied, Bosch switched gears and spent most of his time in the United States courting Standard Oil and Ford. Standard was particularly interested. The company dominated the gasoline business, but was deeply worried about the predicted shortage of crude oil. Bosch's technique promised not only a way to make gasoline from coal but also a way to improve the gasoline yield and quality from crude. With normal refining, it took four barrels of crude to make one barrel of gasoline. With the Farben technology, it was closer to one to one. By using the technique, Standard could stretch its shrinking oil reserves.

Standard Oil was a big company, but even its technical director was stunned by a visit in the spring of 1926 to the company's plants on the Rhine, "plunged," as he put it, "into a world of research and development on a gigantic scale such as I had never seen." He immediately wrote the company's president, Walter Teagle, calling Bosch's synfuels project "tremendously significant—perhaps more than any chemical factor ever introduced into the oil industry." Teagle came to see for himself. After viewing the BASF high-pressure plant at Oppau, he too had to admit how far ahead of the United States the Germans were. "I had not known what research meant until I saw it," he said. "We were babies compared to what they were doing."

Teagle returned the favor in the fall, inviting Bosch to join him on a three-week tour of Standard's far-flung U.S. refineries and oil holdings. When it was over the two men had sized each other up and were ready to work together. Talks with Ford also went well. When Ford decided to start a branch in Germany, Farben bought 40 percent of the stock. When Farben started a branch in the United States, Edsel Ford sat on the board of what was called the American IG. These were just the sort of international ties Bosch wanted.

Another grand machine was beginning to take shape in Bosch's mind. If the world of the twentieth century was to be a prosperous world, a world whose people consumed ever-increasing amounts of what huge industries had to offer, then the leaders of industries with shared interests should simply cooperate. Giant global corporations, in Bosch's mind, should work with each other to rationally fill needs, explore and exploit opportunities, and develop better products. Money spent on competition, in Bosch's view, was money wasted. It would be better used on research. Take his deal making with Ford and Standard Oil: People would not buy Ford cars unless they had access to gasoline; people would not buy Standard gasoline unless they were driving cars; and soon no one would have anything if Farben did not make gasoline out of coal. Of course they should all work together.

There would be other benefits as well. Bosch was also thinking about the future of his nation. Germany had no oil wells of its own. If it was going to become a world power again, Germany would need to join the automotive revolution. Bosch was helping that happen; not only by pioneering a method of production that would fill cars with synthetic gas made in Germany but also by reducing the amount of money flowing out of Germany to pay for imported oil. The saved money could be used for paying reparations. At the same time a steady supply of affordable German gas would help nurture German demand for autos (which was why Ford was interested). Everyone could come out a winner.

Standard Oil thought the same way. In a series of deals signed with Farben in the late 1920s, Standard bought the rights to explore and exploit Farben's high-pressure synfuels process everywhere in the world outside of Germany. In exchange, Standard gave Farben stock worth about 35 million dollars (150 million marks). The two companies created a joint corporation that held title to the patents.

The injection of Standard funds was desperately needed. Bosch's synthetic gas project was in trouble.

FARBEN BEGAN MAKING predictions in the fall of 1926, just as Bosch was touring America. The company announced that Leuna would soon be pumping out a hundred thousand tons of gasoline per year. Six months later, however, only a few barrels had been made. There had been unexpected technical delays, the process was more difficult to tame than they had expected, the coal/oil slurry was chemically complex and physically hard to handle, the list of problems and delays went on. If Bosch had concerns, he pushed them aside. He needed this process to work. He poured money into research, pushed his teams, and ordered his people to move quickly—more quickly than some of them thought wise—toward full-scale production. More than thirteen thousand laborers were put to work rebuilding and expanding Leuna into a synfuels plant. Farben announced that it would meet its hundred-thousand-ton goal by the end of 1927. Taking Farben's line, some experts began predicting that one-fifth of the world's gasoline would be synthetic by as early as 1928.

But not even Bosch's crack research teams could make that happen. By the end of 1927 it was clear that the process would take longer and cost much more to develop than Bosch had predicted. Some directors of the Farben cartel, especially those with roots in companies other than BASF, started to grumble as they saw their income from dyes, drugs, and other chemicals being sucked into the money pit of Leuna.

Bosch was unmoved. He seemed certain that his critics would be proven wrong, just as they had been proven wrong when they wanted to shut down Brunck's indigo project, and just as they had been proven wrong when they wanted to shut down the development of the Haber-Bosch system. There were problems with synthetic gasoline production, yes, but there were bound to be problems during developments of this magnitude. In any case they were being solved, slower than hoped, perhaps, but solved nonetheless and they would continue to be solved.

Then the bottom dropped out. In the late 1920s a new oil field was found in Oklahoma. As oil prospectors swarmed to the area, it quickly became clear that it was bigger than anything they had ever seen. What they had found was only the edge of a huge deposit, an underground lake of oil. No, more than a lake. It was a sea of oil. By 1930 oil wells were sprouting across six states, all the way across Texas and into Louisiana. There were sudden fortunes, the rise of the Texas Oilman, and stories of yesterday's penniless Indians driving Cadillacs. Amid all the excitement came a deeper realization: There was not going to be a worldwide oil shortage. Gasoline, instead of selling for a dollar a gallon (as a U.S. senator had predicted a few years earlier), was as low as nine cents a gallon.

This was a disaster for Farben. All of its financial projections for synthetic gasoline were based on predictions of steadily shrinking supplies of natural crude and steadily rising gasoline prices. In an era of cheap gas, Leuna's product would be too expensive to compete. It was a terrible blow, especially just as Leuna was finally starting to produce. But Bosch and his company were committed now. There was no turning back. They told themselves that even the huge new American oil fields would run dry eventually. Perhaps that would be the end of natural crude. In any case, Germany would still benefit from becoming self-sufficient in oil. Bosch kept talking and kept investing in the synfuels plant at Leuna.

He began looking for more money to finish the project. He had fused together the chemical companies of Germany, in part to fund his new dream for Leuna. He had made what deals he could with international businesses. There was only one more place to go: the government. Bosch had seen how important government funding was for his ammonia process during the war, when deals with the kaiser's ministers had provided the money to build Leuna (and drive competing technologies from the field). Farben representatives increased their political work, bringing politicians to Leuna to impress them with the size and scale of the factory, then planting the idea that support for Leuna gasoline, or *Leunabenzin*, was the same as support for a strong Germany. Every liter they produced themselves was a liter's worth of money kept in Germany rather than spent for oil from other nations. Self-sufficiency in gasoline, they argued, was a national benefit. To make that happen, to allow *Leunabenzin* to compete, all the company wanted was somewhat higher tariffs on imported oil (to raise the price of imported gasoline, and keep *Leunabenzin* competitive within the German market) and some government funding for the plant.

The political wooing continued through the summer of 1929. For the moment, Bosch was comforted by the fact that apart the money flowing to his gamble at Leuna, Farben was healthy enough. Nitrogen sales continued to be surprisingly strong, given the growing competition. Farben might no longer have the only Haber-Bosch plants in the world, but the German company's were arguably the best. Farben was even making a few deals for plants to be built in other countries, in the United States as part of its deal with Standard Oil, and in France and Norway. Bosch was happy to see the company continuing to extend its presence outside Germany, not only building ammonia plants but also extending its operations in the areas of pharmaceuticals and dyes as well, making products in other nations as a way of avoiding foreign tariffs and increasing the company's reputation.

That was good, and there were more things for which Bosch could be thankful. Memories of the hyperinflation and armed labor troubles of the early 1920s were fading. Wages were rising. A large, modern new central administration building for the firm was nearing completion in Frankfurt. Apart from synthetic gasoline, Bosch's heavy emphasis on research and development was paying off with a string of new products and processes, from a popular "mixed" fertilizer designed in his agricultural labs to early steps toward plastics. The most promising line of research was opening what looked like a path to making synthetic rubber. This would be another great advance, and one that promised enormous profits. Rubber was needed for everything from tires to gaskets and hoses, and demand was growing along with the auto industry. Farben's synthetic rubber, in a form they called Buna, was not perfect yet, but it looked like it might soon be. Things could be far worse.

Suddenly they were.

CHAPTER 19

A FEW MONTHS AFTER the thirty-five-million-dollar contract between Farben and Standard was signed the U.S. stock market crashed, kicking off a worldwide Great Depression. It hit Germany especially hard. Ever since the reparations crisis of the early 1920s Germany's government had been propped up with enormous loans from the United States. After the crash, the debts were called in to help assist the faltering U.S. economy. Germany had no way to pay. Income to companies like Farben dwindled as the Depression spread, drying up foreign markets for German goods. Companies started failing, and workers were laid off. The number of unemployed men and women in Germany more than quadrupled between 1928 and 1930, then doubled again during the next two years. Unemployed workers began to listen to the rhetoric of radical political groups on the Left and Right, from communists to Adolf Hitler's ultranationalist Nazis. Germany's political center began to lose power.

Carl Bosch watched as prices plunged, especially for ammonia and other nitrogen products, and his company's income plummeted. No one in the world, it seemed, had much money to spend on gasoline or fertilizer or chemicals, or much of anything else. In 1930, the first year of the Depression, income from Farben's nitrogen processes fell by a third. That made it a relatively good year compared to what came next. By 1933 Farben's income had fallen by half again. As one company history put it, the financial pressure increased from "extremely worrying to nearly unbearable."

Bosch bore much of it. A portrait of him from 1930 shows a thinner, almost haunted face, with dark circles under the eyes. His worries during the early days of the Depression were deepened by yet another threat to Leuna, this time from within his own company. Carl Duisberg, the head of Bayer, the man Bosch had bested for leadership of the company during the Council of the Gods, had generally supported Bosch's policies. In the face of financial disaster, however, the two men broke on the question of Leuna. Duisberg led a faction within the board that wanted to shut down the synthetic gasoline project and end the drain on funds, to put the still-troubled technology on hold at least until the Depression eased. The Leuna critics marshaled their case in a lengthy report on the plant's current progress and future prospects. It was a disaster for Bosch. Completed in 1931, just as the Depression was at its worst, the report estimated that four hundred million marks more would have to be spent to perfect the technology. The only way to cover the costs would be with government loans, and even then the problems would not be over. When running at peak efficiency, the cost per gallon for gasoline from Leuna would still be twice that of a gallon made from Oklahoma crude. They had gone into the project expecting high gas prices. That was how their technology was going to compete. With gasoline selling at historic lows, it began to look, as historian Thomas Parke Hughes wrote, as if Farben had "a vested interest in a white elephant." Duisberg's backers began calling for an immediate shutdown.

Bosch and his right-hand man, Krauch, mounted a furious defense. The company had already invested so much, they said, that the only direction for the synthetic gas project was forward. Shutting down Leuna would throw ten thousand people out of work. Farben would lose the gigantic investment already made in buildings and machinery. This was about more than gasoline. The synthetic fuels project was tied to the profitability of Farben's other processes, most notably ammonia production. Ammonia and synfuels needed each other to stay healthy. Look at hydrogen, they said. Both processes used pure hydrogen in enormous quantities. The more hydrogen Leuna made, the bigger the equipment to make it, the cheaper the unit cost. Expanding hydrogen capacity at Leuna meant lowering production costs for ammonia. Stopping the synthetic gas project meant losing economies of scale, which would cut profits from ammonia— and ammonia and the fertilizer made from it were still the most profitable products Farben produced. If they pulled out, what would happen with the deals with Standard Oil? Don't forget, Bosch argued, that they were just now getting up to speed with the production of *Leunabenzin*, which even at a higher price was beginning to make at least a little money. Plans were in place for *Leunabenzin* gasoline trucks and *Leunabenzin* filling stations. Growing numbers of German-made cars would be running on *Leunabenzin*. The company could pressure the government to raise tariffs on imported oil, making *Leunabenzin* more competitive. It was too late to back out. Bosch and his supporters concluded that shutting down synfuels at Leuna would actually cost more money than moving ahead. Everyone in the company would have to take a deep breath, lower their heads, and push through.

Bosch prevailed, at least for the moment. He managed to avoid a shutdown long enough for Leuna to finally reach its hundred-thousand-tons-per-year gasoline production goal— four years late. It was late, but it was finally working. The political lobbying worked too; by mid-1931 Germany had the highest tariffs in Europe on imported oil. But none of his efforts could reverse the Depression. People had no money to buy *Leunabenzin*. By early 1932 Leuna was operating at just 20 percent of capacity. Bosch saved money by mothballing some sections of the plant and running others at full speed. When one set of machinery wore out, instead of replacing it, he switched to a fresh set. Even now the synthetic gasoline process was not perfect; it still required more fixes, more fine-tuning. There were still good arguments for shutting it down. But Bosch would not. He was like a man under a spell, mesmerized by the great machine he had built. Nothing would make him close Leuna.

THERE WERE TWO bright spots for Bosch early in the Depression. In November 1931 he learned that he had been awarded the Nobel Prize in chemistry for his work on ammonia (he shared the prize with Friedrich Bergius, who won for his discovery of the coal-to-gasoline process). It was a surprise, not because Bosch's work had not been important but rather because he had perfected the industrial application of a discovery instead of making the discovery itself. "Blocky, bristly Professor Bosch," as *Time* magazine described him, a man who "says little, listens much, dresses carelessly, and peers through thick spectacles at the workings of the great machinery he has set in motion," took full part in the ceremonies in Stockholm, dressing formally, shaking the king's hand, and giving a lengthy, somewhat technical acceptance speech.

The other bright spot came in the form of a politician, a man Bosch believed might be able to guide Germany through the Depression without allowing it to fall into chaos. His name was Heinrich Brüning, and he was the nation's new chancellor.

Brüning was the epitome of reasonableness, self-effacing, quiet, studious, and hardworking. Still relatively young—just forty-four years old when he took office in the spring of 1930— he had been a soldier (he won an Iron Cross in World War I), a trade union leader, and a newspaper editor. Most important, he had a doctorate in economics. Some said that Brüning's idea of relaxation was to curl up with a tome on the subject. He intended to pull Germany out of the Depression with firmness, reason, and economic theory. Bosch could relate to a man like that.

Both men understood how precarious the next few years were going to be for Germany; both were political centrists; and both wanted to prove to the world that Germany could govern itself without falling prey to left- or right-wing fanatics. Both believed that international goodwill was a prerequisite for international business. They worked well together.

Bosch once outlined his political views in a speech to a chemists' group, noting that "trust in our leadership is dwindling, and only the call for a strong man who will surely get things right again remains. However, by doing this we forget that today, in the face of huge challenges to be tackled, one single man will never be able to meet the expectations placed in him." On the other hand, neither did he like the communist call for putting individuals second to the welfare of the state. "I stand here in conscious opposition to those who say and demand that the individual will and must subordinate himself to the general good," Bosch said. "I consider such desire absolutely irreconcilable with human nature. Man in his entire makeup and evolution is not a herd animal but a family animal."

Bosch was a liberal technocrat who believed in progress, believed in individual drive, and believed that humans worked best when working freely for themselves and their families. "The purpose of the state," he said, "is to make sure that the gainful employment and co-existence of individuals and nations proceeds with the least possible amount of friction." People should be free to make products, make money, and pursue their interests. Beyond that, government should get out of the way. He would have done well in the United States.

The Brüning regime was good for Farben, setting economic policies that favored Farben's needs (like the high oil tariff) and banning import of foreign nitrogen products. Farben ended up with a virtual monopoly on fertilizer in Germany, which helped them limp through the Depression (although German farmers were unhappy about the artificially high prices). Bosch supported Brüning's political efforts. One of Bosch's board members served in Brüning's cabinet.

Good relations established, Bosch focused on getting Brüning to put money toward Leuna and synthetic gasoline. He needed to buy a little more time, enough to perfect the process, enough to increase *Leunabenzin* production and lower prices just a bit more. Bosch was convinced that it would be in the state's best interest—as well as Farben's—if the government could guarantee some form of subsidy or other support for Leuna and synthetic gasoline. Talks started in the early spring of 1932. But by April Brüning was gone. Perhaps he had been, in a sense, too reasonable for his times. As unemployment continued to rise, Germans began looking for a different kind of leader, a less temperate leader—the sort of "strong man" that Bosch despised.

Germany's spring elections showed that despite Brüning's efforts, the nation was splintering. The intemperate, anti-Semitic leader of the Nazi party, Adolf Hitler, received almost as many votes for president in the election as the winner, the elderly warhorse Paul von Hindenburg.

The growing popularity of Hitler and his Nazis worried Bosch. The Nazis were bad for business. Their inflammatory politics would bring back the anti-German feelings that Bosch and others had tried so hard to ameliorate after the war. Their race-based rhetoric—especially their anti-Semitism—was anathema to Bosch, a man who had spent his career working with Jewish scientists and businessmen. Bosch's personal secretary was the son of a rabbi. The Nazis were happy to mix it up, fighting in the streets with left-wingers, destroying any sense of public unity. They were dangerous. But their strong showing in the

1932 election marked the political end of Brüning. He had managed to avoid Hitler's election as president by working to elect the aging Hindenburg, but soon after the election (partly in an attempt to appease the Nazis) Hindenburg requested Brüning's resignation. Nine months later, after two more chancellors had come and gone in quick succession, Hindenburg appointed a third man. On January 30, 1933, Hitler was made chancellor of Germany. A year later, Brüning fled, first to England, then to the United States, where he taught political science at Harvard.

FRITZ HABER WAS in the middle of a two-month rest cure at Cap Ferrat on the French Riviera in January 1933 when Hitler was named chancellor. Haber's heart had been troubling him again, and he had overworked, and perhaps he wanted an escape from the worsening political news. Hitler's accession did not improve his health. "I am fighting with diminishing strength against my four enemies," he wrote Richard Willstätter, "sleeplessness, the financial demands of my divorced wife, my increasing disquiet over the future, and the feeling of having made serious mistakes in my life." Perhaps he was thinking about his conversion, his attempt to assimilate, and what it meant to be a Jew under Hitler.

He returned to his institute and began to learn to live under the Nazis. Four weeks after Hitler took the chancellor's office, Germany's parliament building was destroyed in a mysterious fire. The nation went into crisis. A young communist was arrested and charged with the crime, and the next day Hitler started suspending civil liberties. Within a few weeks Germany's lawmakers passed an act that allowed the chancellor to take control and make laws as he saw fit. Even the middle-of-the-road parties voted to bestow dictatorial powers on Hitler.

He began using them. In March 1933 state agencies were supplied with new flags featuring the swastika. When Haber's institute was asked to fly one, he forced himself to personally direct the building supervisor in raising it, an act "so much more dignified for [Haber] than if this requirement had been forced upon him," one of his Jewish employees wrote. Perhaps if he flew the flag and kept his head down, the Hitler regime would pass. On April 1, 1933, Nazi leaders called for a nationwide boycott of Jewish businesses, with "Jewish" identified not by the practice of religion but by ancestry. Haber's conversion did not matter. A Jew was a Jew. The same day the justice minister in Prussia demanded that Jewish judges voluntarily step down from their posts. This was particularly worrisome to Haber: Being a judge was one of the highest, most prestigious positions a Jew could hope to achieve in Germany, a symbol of how far German Jews had advanced. Asking them to leave their benches was more than a legal move. It was a preview of things to come. Haber wrote Willstätter that what happened to Jewish judges could well happen to Jewish scientists.

Then, on April 7, 1933, Hitler put into effect a wide-ranging decree designed to cleanse the German government of non-Aryans. The Law for the Restoration of the Professional Civil Service, as it was called, ordered the removal within six months of anyone in government service who was not an Aryan (including Jews) or who criticized the government, regardless of their position or tenure. The only exceptions to be made were for veterans of the Great War (or those whose fathers had served). The exception of war veterans was not Hitler's idea—it was insisted on by Hindenburg, the old general, a figurehead Hitler felt he still needed.

After he was informed of the edict, Haber was too stunned to do anything for a few days. It was as if he had woken up on a new planet. This was the first broad-scale, explicitly anti-Semitic law enacted in Germany during his lifetime. It was a repudiation of everything he had hoped his nation was coming to be. It was the end of his German dream.

CHAPTER 20

HITLER'S EDICT WAS announced during Germany's long Easter break, when many academics, including Fritz Haber's institute staff, were on holiday. The institute would not reopen fully until the end of the month. His office received questionnaires to be filled out by every employee, including Haber, asking for details on family ancestry. About a quarter of his people, including two of his department heads and many of his top researchers, were Jewish. There were uncertainties about the interpretation of the decree, about exactly how decisions would be made and implemented. The exception for veterans seemed to provide a way out for Haber and some of his older scientists, including those who had worked with him on poison gas. But how could he stay on when many of the people who worked at his institute would be forced out simply for being Jewish? How could he bring himself to sign the dismissal papers?

While he thought about options, others made decisions. Albert Einstein, traveling in the United States, attacked the Nazis in print and never returned to Berlin. James Franck, a Jewish chemist who had worked on the front lines with Haber on the poison gas project, announced on April 15 that he was resigning his professorship at Göttingen rather than be forced out. "I can't just get up in front of my students and act as though all this doesn't matter to me," he wrote Haber. "And I also can't gnaw on the bone that the government tosses to Jewish war veterans. I honor and understand the position of those who want to hold out in their positions, but there also have to be people like me. So don't scold your James Franck, who loves you."

Haber was torn. Part of him hoped that the madness would blow over, that the German people would not stand for it, that Hitler's government would see sense. Didn't the Nazis realize that forcing out Jewish researchers would gut Germany's science programs? Jewish scientists were simply too successful, too integral to Germany's international eminence. How could they be told to leave without doing irreparable damage? How could the Nazis throw away a century of achievement? It was beyond belief. It was madness.

He then was notified by the Nazi authorities that his institute could not reopen after the spring holiday without first reorganizing along the lines demanded by the civil service edict. The Kaiser Wilhelm Institutes in general and Haber's institute in particular seemed to be targeted by the Nazis. "The founding of the Kaiser Wilhelm Institutes in Dahlem was the prelude to an influx of Jews into the physical sciences," the German national student newspaper for the sciences wrote around that time. "The directorship of the Kaiser Wilhelm Institute for Physical and Electrochemistry was given to the Jew F. Haber, the nephew of the big-time Jewish profiteer Koppel. The work was reserved almost exclusively to Jews." (Haber was not related to Koppel and three-quarters of the workers in his institute were not Jewish.)

Other leaders within the KWI advised Haber to go along for the moment, ride out the storm, keep his post and let a few of his Jewish researchers go in order to appease the authorities. Still stunned, still unsure of what to do, he followed their advice. Among the first to leave were his two Jewish department heads, highly respected men whom Haber felt would quickly get jobs outside of Germany (they did, in England). It was not good enough for the Nazis. They demanded a complete cleansing, with only the veterans like Haber excepted.

As he looked over the lists of others he would have to dismiss, men and women he had hired, who had done excellent work, something turned in him like a key in a rusty lock. He decided to resign. On April 30 Haber wrote a letter to the Prussian Ministry of Education. A good civil servant to the end, he buried his emotions in bureaucratic prose: "My decision to request retirement derives from the contrast between the research tradition in which I have lived up to now and the changed views which you, Minister, and your ministry advocate as representatives of the current large national movement," he wrote. "My tradition requires that in a scientific post, when choosing co-workers, I consider only the professional and personal qualifications of applicants, without considering their racial make-up. You will not expect a man in his 65th year to change a manner of thinking which has guided him for the past thirty-nine years of his life in higher education, and you will understand that the pride with which he has served his German homeland his whole life long now dictates this requirement for resignation." He asked only that he be allowed to stay in his post until October, to clean up the institute's business, to find a successor, and (he hoped privately) to assist as many of his Jewish staff as possible find new positions.

Haber's resignation created headlines around Germany. It was not necessarily news that the Nazis wanted. Einstein's displeasure with the Hitler regime meant nothing: Einstein was no patriot, but an eccentric socialist/pacifist, a loudmouth who abandoned Germany in favor of America. Einstein could be demonized and dismissed. But Haber was a different matter. Haber was Germany's foremost chemist, renowned for his service to the state as well as his research. His resignation caused a furor, both among the Nazis and among those German scientists who now realized that Hitler's new law might mean the end of Germany's preeminence in science.

Haber no doubt received private support and condolence. But the times were such that no one dared to say much in public. During the late spring and summer Max Planck, president of the Kaiser Wilhelm Society, worked the government agencies, trying to reach a compromise that might convince Haber to keep his position. The first response came from the Nazi culture minister, who told Planck simply, "I'm finished with the Jew Haber." In May Planck took his case directly to Hitler, visiting the chancellor in his Berlin offices, hoping to make the rational case that forcing Jewish scientists like Haber to flee Germany amounted to German scientific "self-mutilation." No transcript of the conversation exists, but Planck remembered Hitler saying that he was more concerned with communists than Jews. The problem, Hitler explained, was that "Jews are all communists. A Jew is a Jew. . . . They all cling together like burrs." The only answer to the problem was to proceed against them all. When Planck tried to return to his point about science, the chancellor began talking faster and louder, pounding his hand on his knee, flying into a rage so fierce that the elderly Planck had to leave the room. Some time passed before he recovered emotionally.

No compromise was possible. Haber presided over the destruction of his own institute, doing what he could to ease the way for his fired Jewish co-workers—his employees joined an exodus of Jewish university workers across Germany, from secretaries to professors—while trying to marshal his own flagging energies. There were no significant public protests from students or colleagues. German university students were, in general, ready to shed the stigma of World War I, eager to free their nation from the shackles of Versailles, and devoted to making Germany great again. They were strongly pro-Nazi. Among faculty members, there was great interest in seeing who would get the vacant positions.

Haber was shattered. It was not just his health this time, not just nervous exhaustion, but a feeling that he created a life for himself that was false—that he had been living a lie. He had spent his career fashioning himself into the perfect German. He now understood what that meant—in Hitler's terms. Despite his conversion, his Nobel Prize, his Iron Cross, his efforts to save the nation, his international stature, his service, his achievements, and his value as a man, the only thing that mattered was that he was a Jew. Einstein, writing to his old friend Haber from the United States, put Haber's dilemma into scientific terms. "I can imagine your inner conflicts," he wrote Haber. "It is somewhat like having to abandon a theory on which you have worked your whole life. It's not the same for me because I never believed in it in the least."

Haber's response was more emotional. "I am bitter as never before, and the feeling that this is unbearable increases by the day," he wrote his friend Richard Willstätter. "I was German to an extent that I feel fully only now, and I'm filled with incredible disgust."

CARL BOSCH'S FIRM was not a state agency and the fate of his employees was not determined by the civil service edict. But he also knew that the Nazis thought of Farben as a tool of Jewish international capital, and called it "Moloch IG" in their propaganda. They had taken a stand against animal testing (likening it somehow to Jewish ritual slaughter) and seemed in general to value hazy Teutonic myths over scientific facts. But they attracted votes and they looked like they were going to gain power. When faced with a number of candidates, it was better corporate policy for Farben to support everybody a little and then support the winner a lot. During the early 1930s Bosch did his best to hold his nose and put up with Hitler.

Farben representatives started to make contact with the Nazis after the Brüning government fell and took the political center with it. As early as November 1932 Hitler met with some of Bosch's people to talk about the future of synthetic gasoline. The meeting—scheduled to last a half hour but running more than two—was fateful. Hitler, it turned out, was a fan of the automobile. He saw a future in which the German people drove German "people's cars," (*Volkswagens*) on high-speed German highways (*Autobahns*). He knew they would need gasoline, and he knew that Germany had no natural source—except what Farben could make. He was surprisingly well informed about the technical side of the Bergius process. He was in favor of Leuna, he told the Farben men, adding that "German motor fuel must become a reality even if this entails sacrifices." It was everything they could have hoped for. When his executives told him about the meeting, Bosch is said to have remarked, "The man is more sensible than I thought."

On February 20, 1933, a week before the Reichstag fire, a secret meeting was held in Berlin between high-level Nazis and a group of two dozen leading German industrialists and bankers. Farben sent high-level representatives (Bosch himself did not attend). Spring elections were just a few days away, and it looked as if the popular vote was going to swing even more to the Nazis, solidifying Hitler's hold on power. The businessmen expected to hear a sales pitch from one of Hitler's top men, Hermann Göring, along with a request for political donations. There was a murmur in the room when Hitler himself entered, shook hands all around, and took a seat at the head of the table. It was the first time many of the businessmen had seen him in person and the first detailed exposure many of them had to the new regime's political and economic plans. Hitler spoke for an hour and a half, speaking persuasively and without notes, describing in general terms his plans for pulling Germany out of the Depression, attacking the communists—a message that played well with his capitalist audience—insisting that "only a martial nation can have a flourishing economy," and mentioning plans to resuscitate the mighty Germany army. He noted the importance of his party's overwhelming success in the upcoming elections and hinted that if the Nazis did not achieve control democratically, they might have to take to the streets and seize it. Then he rose and took his leave. His representatives asked the attendees to contribute generously to the Nazi Party. Goals were set for each industry. Chemical producers were expected to chip in five hundred thousand marks each. Bosch's people took notes, shook hands, and later made their report. When Bosch heard about Hitler's speech and the funding request, his only response was nonverbal: He shrugged his shoulders.

It was a gesture of resignation. It was becoming clear that the Nazis were going to win the elections, the other companies were already donating large sums, and it was time for Farben to smooth relations with the government in power. Perhaps, he hoped, Hitler would be more sensible than he appeared. Whatever happened, he was likely to be chancellor for at least a few more years. Bosch needed government money for Leuna. A few days after the meeting, Farben deposited four hundred thousand marks into a political fund. Almost all of it went to the Nazi Party. Another hundred thousand marks was given to a more moderate political group. The total was less than Farben usually gave during an election year, but it was an important gesture. The Nazis won a majority in the spring elections.

Bosch lost no time getting a return on his investment. He quickly opened talks with Hitler's Ministry of Economics to secure massive Nazi support for synthetic gas. The minister seemed to understand the need for German fuel independence. It looked as if things might move in Farben's favor.

Then Bosch made the mistake of meeting with Hitler privately.

It happened within weeks of Haber's resignation. Hitler was forming a high-level economics council to provide advice to his government and had asked Bosch to serve. Some of the prospective members, including Bosch, were asked to private interviews in Berlin. Bosch had been carefully tracking the Nazi's economic policies and noting Hitler's increasingly radical pronouncements. He had been stunned by the purge of Jews from the civil service. The chancellor seemed capable of veering suddenly from sensible to dangerous. Hitler, for his part, had his secret police gather information on Bosch's political views. They found that he was a typical southwest German liberal, a highly educated friend of the Jews, but a man who seemed to put business and science above ideology. Bosch tended to complain about the Nazis—he had felt compelled to make his views about the civil service law known to members of the Kaiser Wilhelm Institutes, for instance—and appeared to want to keep his own Jewish scientists in place at Farben. The secret police knew that he had no intention of joining the Nazi Party. They knew that he had, in 1929, bailed out a foundering liberal newspaper in Frankfurt. They knew that he was friends with Hermann Bücher, the head of a large German electrical firm, a man so vehemently anti-Nazi that friends had to warn him to keep quiet in public. The newly formed Gestapo was in the process of gathering large dossiers on both Bücher and Bosch. (Luckily someone soon warned Bücher about his secret file, and he somehow managed to have both his and Bosch's stolen from the Gestapo offices. He was said to have burned them in his home fireplace.)

Bosch's meeting with Hitler started well enough. Bosch talked about synthetic gasoline and the need to expand Leuna. Hitler seemed agreeable. Then Bosch felt compelled to start talking about the civil service law and the damage that would be done to German chemistry and physics if it was applied unsparingly to Jewish scientists. Hitler lost his temper. He shouted, "You don't understand these matters!" and started ranting about the Jewish threat. If Jews were so important to physics and chemistry, Hitler said, "Then we'll just have to work one hundred years without physics and chemistry!" When Bosch tried to disagree, Hitler, in a calculated insult, rang for an aide and announced that his visitor wished to leave.

Bosch was stunned. He told friends later that Hitler seemed to go into a sort of trance when he was excited, like a man lost in a dream. The Farben staff did its best after that to keep Bosch away from Hitler, and the two men never spoke privately again. They were in the same room together only once, when Bosch attended the first meeting of Hitler's council on economics. Bosch, who had been asked to give a speech, planned to say something about the benefits of open, international business and scientific communication. He knew that the topics were risky, given that Hitler would be in the audience, but, he said, "People should not have to say later on, 'Oh, if only you had opened your mouth.'" When Hitler learned that Bosch was going to speak, he turned and left the hall. The Nazi organizers ordered that a new speaker be found. When the group declined, they closed the meeting. The economics council never met again.

BOSCH NOW BETTER understood what Hitler represented. He thought about Farben's many Jewish researchers and board members. He thought about Haber, who at that time was still struggling with the dissolution of his institute. They had never been close during the years following their work on ammonia, but now, following Haber's resignation and his own "interview" with Hitler, Bosch felt a sort of kinship. He dictated a note.

"I heard with great regret in Berlin how very oppressed you feel personally by the present circumstances," he wrote Haber. "You might know that I myself have tried everything possible in order to make the measures against scientists somewhat bearable; and I do not need to assure you that the personal side of the whole movement affects me extremely deeply. . . . If I can be of any assistance to you somehow, then I am naturally gladly at your disposal."

Haber was happy to receive the note. It was the first communication he had gotten from anyone at Farben. He wrote back, telling Bosch that he saw his retirement "not as a hardship but rather as a relief," thanking him for his good wishes, and letting Bosch know that he was looking for a position in another country. "I believe your readiness to come to my aid if I were to ask you," he ended. "But I do not know what I could ask for, unless it were that you refuse to listen to those people who, subsequent to unfriendly newspaper portrayals, pass me off as a sick old man."

Haber began searching for a scientific position for himself, putting out feelers in England, Holland, Sweden, Spain, Paris, and Palestine. Offers came in more slowly than Haber might have hoped. He was of retirement age, his heart was failing, and he was more accustomed to administrating than working in a laboratory. That, and his lingering reputation as a war criminal, all served to cool any enthusiasm there might have been to shelter him.

Haber was not helping his own case: He seemed unable to focus. Sometimes he seemed to think that the best course might be to stay on in Germany. The next moment he wanted to leave immediately. His job hunt led to talks with Jewish chemist and Zionist activist Chaim Weizmann—later the first president of Israel—who tried to lure Haber to Palestine with offers of a position in a new scientific center. Weizmann wrote that he initially found Haber "lacking in any Jewish self-respect," a man who had "converted to Christianity and pulled his family with him along the road to apostasy." But he came to appreciate the depth of Haber's confusion and his need to find his way back to faith. Haber began writing to others, including Einstein, about his newfound enthusiasm for a Palestinian homeland.

Haber's position at Dahlem did not end officially until October 1933, and he seemed intent on staying despite his failing health. In August he took a long-planned trip to Spain to speak at a chemical conference—a great strain, considering his worsening health. On the way he visited his son Hermann, Clara's boy, now married and living in Paris. His old friend Richard Willstätter visited him there and was shocked to see how sick Haber looked. In Paris Haber finally received an acceptable offer of a scientific position, a laboratory and nonteaching faculty position at Cambridge in England. He was being given a chance to make a new home in a former enemy nation, and he wrote back accepting the offer. He was especially interested in the chance to gain British citizenship. "My most important goals in life are that I not die as a German citizen," he wrote, "and that I not bequeath to my children and grandchildren the civil rights of second-class citizenship, as German law now demands that they accept and endure on account of their Jewish grandparents and great-grandparents." The perfect German was ready to become an Englishman. Then he continued to his meeting in Spain, stayed up all night preparing a scientific talk, and the next day—gasping, trembling, and taking nitroglycerin to ease the pain of a heart spasm—delivered it.

As he traveled back to Germany on the train, Haber found himself in distress, his heart stuttering, his mind racing, unable for the first time in his life to properly arrange all the little boxes that he had made: one for his Jewishness, another for his Germanness; one for his family, another for his work; one the genial scientist, another the secretive schemer; one the public benefactor, the other obsessed with glory and money. The boxes began to crack open and empty into one another. Haber was flooded with conflicting emotions. For once he could not make up his mind. "I was German to a degree which no one today would believe," he said, and yet he had accepted a position in a former enemy nation. There was no way of telling how long that position would last. Part of him still dreamed of reclaiming some semblance of his old life in Germany. He was still being courted by Chaim Weizmann for a position in Palestine and had not found it within himself to close that door. Einstein had written him again. "I'm especially glad . . . that your earlier love for the blond beast has cooled off a bit," Einstein wrote. "Who would have thought that my dear Haber would appear before me as an advocate of the Jewish—and even Palestine's—cause!" Einstein thought Haber should never return to Germany, where, he wrote, so-called intellectuals had been shown to be "men who lie on their bellies before common criminals, and even sympathize with those criminals to a certain extent." All this was running through his head as his train raced toward Germany.

He got as far as Basel, Switzerland, a few miles short of the border, when his health took such a turn for the worse that he had to be carried off the train. He was still alive, but he was too exhausted and sick and weak and old to continue. His heart would not let him.

After resting enough to regain his ability to walk, he learned that Weizmann was nearby, staying in Zermatt, taking in the views of the Matterhorn. His physician warned him not to travel to any altitude—Zermatt was an alpine resort town, almost a mile high—but Haber made the pilgrimage anyway. He arrived with enough energy to launch into a dinnertime speech about the importance of a Jewish homeland.

During the course of the evening, Weizmann remembered, Haber also spoke about himself. "I was one of the mightiest men in Germany," he said. "I was more than a great army commander, more than a captain of industry. I was the founder of industries; my work was essential for the economic and military expansion of Germany. All doors were open for me." That now meant nothing. "At the end of my life I find myself a bankrupt," Haber said.

The important thing, the two men agreed, was a Jewish homeland. Weizmann repeated his offer of a scientific position in Palestine, noting that the climate would be good for Haber's health. "You will work in peace and honor," he said. "It will be a return home to you—your journey's end." According to Weizmann, Haber agreed to take the position at some time in the future, perhaps after he saw what life held for him in England.

On his way down the mountain, Haber suffered a major heart attack.

HE WAS STILL alive, but barely. It took several weeks at a sanatorium on Lake Constance before Haber felt well enough to travel again. September turned into October and the date for his official departure from Dahlem came and went, marked only by a note from Haber that was posted on his institute's bulletin board: "With these words I depart from the Kaiser Wilhelm Institute . . . which under my leadership for 22 years was dedicated to serving humanity in times of peace, and the fatherland in times of war. As far as I can evaluate the result, it was good and useful for science and for the nation's defense."

It was the end of his career in Germany. He now needed full-time care, provided by one of his sisters, who attended him in Switzerland and helped him dispose of his furnishings and business matters in Berlin. By mid-October Haber was homeless. He had enough strength to give England a try, traveling to Cambridge at the end of October, taking rooms for himself and his sister in the University Arms Hotel. He was given a university laboratory, a German-speaking assistant, and a problem to work on—the catalytic decomposition of hydrogen peroxide. He made an attempt to settle. It was a happy time in some ways. Just before Christmas, a group of co-workers showed up, including a number of laboratory assistants from his Kaiser Wilhelm Institute, and held an informal Haber colloquium in his hotel room. "Then began a scientific discussion more wonderful than you could possibly imagine," one of them wrote. "All cares, all difficulties, all pressures were forgotten in that moment. And so the Dahlem circle arose anew under Haber's influence in Cambridge, unfortunately only for a short time."

For the most part, however, Haber appeared "ill, depressed, lonely, a shadow of his former self" as one German visitor noted. His English hosts seemed welcoming enough, but some of the laboratory technicians, men who had fought in the trenches or knew those who did, avoided him. The great English physicist Ernest Rutherford refused to meet him because of his war activities. He felt like an outcast. "I fear that I didn't adequately realize what it means, at my age, to move into a foreign language and lifestyle," he wrote Weizmann. He was invited to stay at Cambridge for as long as he liked, but as the dreary English fall and winter went on he became increasingly unhappy and restless. He longed for some warmth. He wanted the sun. He began to plan another trip, a long one, this time to Palestine. All he needed was some money to make his life there comfortable. Haber, during his final weeks, was once again consumed with money worries—the German government had taken a good piece of his fortune in the form of a steep emigration tax—and had decided, toward the end, to reach out to his old co-worker Bosch for help.

The letter was received in the Farben offices toward the end of 1933. "You are now a man of decisive importance," Haber wrote. "You offered me your aid, of your own accord, and I took your words seriously. . . . Won't you make it possible for me to live out these remaining years of pitifully diminished health and strength in peace and decency?"

There is no reply on record.

●   ●   ●

HABER WAS WEAKER by the day, in pain, yet seemed resigned to it. Before leaving England he suffered another mild heart attack in his hotel room, but by then he was so accustomed to these spells of severe ill health that he did not even bother calling a physician. His doctors told him in no uncertain terms that he was not to attempt the Palestine trip. But in January 1934, after spending only two months at Cambridge, he went anyway.

He succeeded once again in getting only as far as Switzerland before he was forced to rest. His personal physician and his son Hermann hurried to join him in his Basel hotel. The physician said later that Haber could not talk for more than a few minutes without having a severe heart attack. He rallied for a few hours that evening, talked with his son, and then went to bed. Almost immediately, he called his physician to his room. This time it was very bad. A local heart specialist was called in and both doctors did what they could for Haber, but his heart was too damaged. He died that evening.

Haber had written a will in Cambridge saying that he wanted his ashes to be buried in Dahlem, next to Clara's. If anti-Jewish sentiment in Germany made that impossible, he wrote, Hermann should decide where they were to be buried. The only thing he stipulated unequivocally about the final resting place was that he and Clara be together. He suggested that the inscription read simply, "He served his country in war and peace as long as was granted him."

Hermann buried his remains in Switzerland. It was 1937 before he was finally able to get his mother's ashes out of Germany; he placed them next to Haber's. On their gravestone he inscribed only his father's and mother's names, with their dates of birth and death. He could not bring himself to add anything about his father's service to Germany.

CHAPTER 21

WHEN CARL BOSCH learned of Fritz Haber's death, he may have regretted not writing him back, but he had been busy navigating the new political waters in Germany, helping his own Jewish employees, and trying to seal a synthetic gasoline deal with the Nazis. There were many blows to bear in these new days, and Haber's death was just one more. Yes, Bosch owed his career to Haber's discovery. But his attention now was focused not on the past, on things that could not be changed, but on his company's future in the new Germany.

After the civil service decree and his talk with Hitler made plain the government's intentions, Bosch toned down the public display of his personal anti-Nazi feelings and concentrated on helping his own people. Public displays of support were encouraged by the Hitler regime; the directorate of Farben attended a May Day rally at the dye plant in Ludwigshafen in 1933, a display of allegiance to the new government replete with swastika flags, goose-stepping troops, and thrice-repeated *"Seig heils!"* Farben paid the Nazis more than lip service, depositing nearly one million marks in a "Hitler Fund." Now that the Nazis were firmly in control, the purse strings were loosened and money flowed from the company to the government in increasing amounts.

At the same time, Bosch did what he could to protect his Jewish employees. Farben was a private business and not subject to Hitler's civil service decree, but the government move had made it acceptable to publicly display a hatred for Jews, and all the long-simmering resentments—Jews making money and taking power, Jews somehow causing the Depression, Jews selling out Germany at Versailles—boiled over. In 1933 and 1934 there were beatings, book burnings, broken shop windows, scrawled insults, and death threats. The Nazi Party in Nuremberg published a cartoon strip called the *Life and Deeds of Isidor G. Faerber*, featuring crudely stereotyped Jewish professors selling substances harmful to the Aryan race. Pressure grew on German companies to dismiss Jewish employees. Many firms were happy to oblige. Others, like Farben, moved only grudgingly. Often they did not have to do anything. Those Jews who understood what was happening were already doing their best to get out, quitting their jobs and fleeing Germany by the tens of thousands. A census in June 1933 showed that there were somewhere between five hundred thousand and six hundred thousand Jews in Germany (comprising only about 1 percent of the general population, although according to one estimate about 20 percent of all German scientists were Jewish); around fifty thousand of them left the country soon after Hitler came to power. Four of Farben's nine Jewish board members resigned early in 1933. Many of its Jewish scientists and executives were gone by the end of 1934.

Bosch succeeded in reassigning some of his top Jewish employees to positions in the United States or Switzerland, places where he thought they would be safe. At the same time he began searching for a compromise, a way to accommodate the Nazi concerns about Jews without forcing them out of the country. Bosch began drafting a plan with other leading businessmen, an alternative to Hitler's policies in which Jews would be accepted in Germany but would be shifted toward less visible, less powerful positions, perhaps moved to rural areas. That done, each "patriotic non-Aryan," according to the plan, would "enjoy the same rank and the same respect as every man." These ideas were never formally submitted to the government. Events were moving too fast; before it could be hammered out, it was clear that no compromise would be possible.

Through it all, he remained committed to synthetic gasoline and to Leuna. In the summer of 1933 Carl Krauch had finished a major report for the Nazis on the need for *Leunabenzin* as a central part of a national fuel program, and it had been warmly received. Negotiations were started on a deal that would bring massive government support to the synthetic fuels enterprise. It was signed in December 1933, as Haber was living out his last weeks. Bosch got almost everything he wanted for Leuna. The Farben board had been worried about the high prices the company would have to charge for *Leunabenzin*, and the weak sales that would result. The December agreement ensured that Hitler's government would buy every gallon of synthetic fuel made at Farben (above what sold on the open market) at a price high enough to cover all production costs—plus a bit of profit. In exchange, the Nazis asked Farben to expand the plant and boost production. They wanted all the gas they could get. Hitler had plans for it. Bosch promised to triple Leuna's production to somewhere around a thousand tons of gasoline per day (about a third of a million tons per year) by 1935. It seemed like a no-lose situation. The Nazis were guaranteeing sales, guaranteeing a price, and guaranteeing a vastly increased market for Bosch's gasoline. Bosch was guaranteeing a close association between Farben and new regime.

There might have been other, less-public agreements as well. Just before the deal was signed, Bosch wrote a widely reprinted newspaper essay titled "Where There's a Will, There's Also a Way." (The use of "Will" in the title is an implicit tie to the Hitler government, as it was in the film *Triumph of the Will*, recorded the next year.) In it, he expressed his growing confidence in the health of the rapidly reviving German economy. "The reason for this I see in the fact that for the first time since the war a German government doesn't just promise," he wrote, "but acts." He praised the Nazis for creating new jobs, lowering taxes, rooting out communists, and restoring pride in German labor. It was not exactly a reversal of everything he had been saying in private, but it was close.

As soon as the deal was signed in December, Bosch ordered another grand expansion for Leuna.

It appeared that, in essence, Bosch was getting into bed with the Nazis in exchange for keeping his dream factory alive. Leuna's health seemed to drive him. He seemed in some ways obsessed with the plant as a physical expression of his personal vision, the symbol of his rise to power, the big gamble he intended winning, his signature achievement, the possible savior of his company, and the machine-city that pointed the way to the future. To save Leuna, it seemed he was prepared to make a deal with the devil. As with all such deals, the real price would be disclosed and paid only later.

MAX PLANCK WAS seventy-six years old, the grand old man of physics, winner of the Nobel Prize, head of the Kaiser Wilhelm Institutes, and he had seen enough. He had tried his best to keep the Nazis happy while ensuring at least some degree of scientific integrity for his fellow Germans, Jews and Gentiles alike. But he, like Bosch, did not like it. A fellow physicist, Paul Ewald, remembered Planck opening a KWI event around 1934, when every official function was supposed to begin with a Nazi salute. "Well, Planck stood on the rostrum and lifted his hand half high, and let it sink again. He did it a second time," Ewald said. "Then finally the hand came up and he said, '*Heil Hitler.*' It was the only thing you could do."

Now, however, at the end of 1934, he felt he had to do a little more. The issue was Haber's ghost. It had been nearly a year since his death and little had been said to memorialize Germany's greatest chemist. The newspapers, knowing that Haber had left his nation in disgrace, hardly noted his passing. Memorial speeches from fellow scientists had been few and scattered. Given Haber's achievements and stature, it was insulting.

The only top-tier researcher to break the silence had been Planck's friend and ardent anti-Nazi Max von Laue, a Nobelist himself, who gave a moving elegy honoring Haber. Laue was fearless—sometimes it seemed he was the only scientist in Germany with enough courage to say exactly what he thought, all the time. Perhaps Laue's courage infected Planck. In any case, the old man decided in mid-January 1935 that the KWI would sponsor an anniversary memorial for Haber, a large public event attended by the top scientists in Germany. It was an unusual move, a combination of a salute to a great German-Jewish scientist and a slap at the Nazi bureaucracy. Planck personally arranged the invitations, the event to be held at Dahlem, on January 29, 1935, one year to the day after Haber's death.

When Planck's plan came to the attention of the Ministry of Education, an edict was sent immediately, forbidding attendance by any German state employee. "Professor Doctor Haber was released from his office on the basis of a request in which his inner attitude against the present state was unequivocally expressed," the ministry stated, "and in which the entire public had to see a critique of the measures of the National Socialist state." There was something odd about the timing too. The planned Haber memorial would fall just one day before the anniversary of Hitler's appointment as chancellor, January 30, a day of nationwide Nazi celebration. An event honoring a Jew in such close proximity was doubly unacceptable.

The Minister of Education's edict was soon echoed by other ministers. It looked as if no one might be able to come. Planck protested that this was a proper event, that he and his people were loyal to the state, that the invitations had already been sent, and that attendance should be allowed. In essence, the old man challenged the Nazis to forbid the event itself. They chose not to do so. State employees would certainly not attend in the face of the official ban—the state employees would include most university professors, most of Haber's colleagues—and with attendance lowered, perhaps the event would be a public failure. The world would see how few Germans cared about a dead Jew.

Planck forged ahead, arranging a hall for five hundred people and putting together an appropriate speaker's list (some who wanted to speak could not attend because of the ban). On the day of the event, he stood in the hall and waited to see who would show up. Nazi functionaries stood near the door, ready to take the names of everyone who entered. People began trickling in. Many of them were women, the wives of men who could not be there themselves because of the government ban. They came to represent the feelings of their husbands. Some military men filed in, colleagues of Haber's from the Great War, a few in uniform. The room began to fill. Then, led by Carl Bosch, a large contingent of men and women from IG Farben arrived and packed the hall. When Bosch had heard about the event from Planck, he spread the word, personally urging chemists and engineers from the old BASF operation to honor the man who had made their fortunes, and telegramming invitations to all his directors. His office helped arrange travel for the trip across Germany from Ludwigshafen to Berlin. He had, it appeared, sold only part of his soul to the Nazis, or kept his fingers crossed when the deal was signed. It seems more likely that he kept making anti-Nazi gestures reflexively, like a fish fully hooked and being hauled in, giving a few last tugs on the line.

There was standing room only when Planck rose to open the memorial. He gave the obligatory Nazi salute, made introductory comments, and gave the podium to a military officer who praised Haber's wartime service. Otto Hahn, director of the Kaiser Wilhelm Institute for Chemistry, honored Haber's scientific work and warmly recalled his personality as a leader of science.

The Haber memorial was not a revolutionary event. It was only a gesture. But it was the only time after Hitler's accession to power that German scientists and their representatives would gather publicly in defiance of Nazi displeasure.

The German Ministry for Public Enlightenment and Propaganda made sure that no newspaper reports of the event were published.

THE MEMORIAL SEEMED to have emboldened Bosch. A week later he sent a memorandum to the minister of education defending freedom of inquiry and noting the importance of scientific research done without thought of immediate utilitarian gain—a direct response to Nazi calls for directing all scientific and business research toward state goals. Shortly after, he attended a national research meeting and made his views about the Nazis clear. A visitor from Stockholm noted that Bosch "participated most actively in the discussion and in such a forthright manner, with so little diplomacy, that I was as amazed as I was pleased." Another participant later told Alwin Mittasch, "We all held our breath. But [the Nazi representative at the meeting] remained calm; he probably didn't even understand what Bosch had said or meant."

Bosch was walking a tightrope. The Hitler regime was making it clear that private interests mattered less than the public good, that all German industries were expected to support the state unreservedly, and that those that lagged would suffer. At the same time, Hitler had quickly defeated the Depression and seemed to be leading Germany, finally, into an era of stability and plenty. Bosch could not expect to continue making deals with the Nazis while at the same time attacking them. The

Farben board realized that even if Bosch did not. In addition to synthetic fuels, it was also negotiating support for the development of Farben's Buna artificial rubber, another big gamble, another raw material important to the Nazis.

Bosch was becoming a loose cannon, and in 1935, the board began tying him down. Bosch was just over sixty years old, too young to retire, so following the death of Carl Duisberg in March, the board moved him out of day-to-day executive decision making and into Duisberg's vacant position as head of Farben's managing board. This was, on paper, a promotion (the managing board, elected by the stockholders, appointed the executives like Bosch who ran the company) but in fact put Bosch into something more of an honorary post, one step removed from research decisions and day-to-day management. It was a move that rewarded Bosch's service while muting his influence. Bosch's former job as top executive went to one of his longtime lieutenants, the buttoned-down and far more politic Hermann Schmitz, a numbers man who rose through the financial side instead of the science side. In important ways, Bosch was stripped of active leadership of the company. Without Bosch making waves, the firm was free to bolster its relationship with the Nazis. The move coincided with Hitler's revelation that he was no longer abiding by the terms of the Treaty of Versailles and that he had secretly rearmed Germany during the past two years, tripling the size of its army and building a twenty-five-hundred-plane modern air force. With Bosch moved aside, Farben was free to supply synthetic fuel for Hitler's new air force and to make artificial rubber for the tires of Hitler's fast-growing land forces.

It is not surprising that after 1935, some of the fight went out of Bosch. He started drinking in earnest. He had always enjoyed alcohol; it provided him with a safety valve, a fast way to decompress. Now it increasingly provided him with an escape. Those around him began noticing that his occasional bouts of depression were becoming more frequent and lasting longer. Some people said he started taking painkillers. He came to the offices in Frankfurt less frequently, holed up in his Heidelberg villa more, took fewer meetings, and slowly became something of a solitary, tended by Else, surrounded by his collections, staying up all night and snapping photographs of the stars.

But there was to be one more professional move. When the aging Planck retired as head of the Kaiser Wilhelm Institutes, Bosch, in 1937, was chosen to replace him. His attitude toward the Nazis had not changed. During negotiations for the position he asked specifically about the organization's policies regarding non-Aryan employees, and the KWI officials assured him that it had not adopted the Nazi line. Bosch agreed to the post. Then, in the institutes' last meeting prior to his assumption of duties, the directors made an about-face and officially adopted the Nazi principles. Several Jewish employees immediately resigned. It is unclear whether Bosch was informed before he took office.

By then, Farben was on its way to becoming completely Nazified. The process of tying the firm's future to Hitler, starting with Bosch's synthetic gasoline deal, was culminating in something like a complete fusion of company and state. The last remaining Jewish officials were ushered out. In a meeting in Bosch's home in April 1938, a Farben committee decided to dismiss all non-Aryan employees at every level. Farben was not alone—the Nazis demanded that all industries toe the line—but Farben was especially valuable to Hitler. Thanks to Bosch's deal, it was responsible for fueling his military ambitions.

Back in 1932, just before Hitler first took power, Bosch had said about the Haber-Bosch ammonia process, "I have often asked myself whether it would have been better if we had not succeeded. The war perhaps would have ended sooner with less misery and on better terms. Gentlemen, these questions are all useless. Progress in science and technology cannot be stopped.

They are in many ways akin to art. One can persuade the one to halt as little as the others. They drive the people who are born for them to activity."

He had been driven to activity, and the consequences were now clear. His friend and fellow anti-Nazi Hermann Bücher, watching in despair as Bosch sank into depression, wrote, "During the years before his death, it became an *idée fixe* that it was he himself who, without wanting it, had made Hitler's policies possible." That was accurate. Bosch's life work, his breakthroughs and factories, his attempts to feed the world and make profits for his company, were being used to arm and fuel the Nazi machine.

In the fall of 1938 Bosch was visited by the commander in chief of the German armed forces and the chief of the army. They were concerned about Hitler's secret plans to attack Czechoslovakia. They thought that the führer might be moving too soon, that the army might not be well enough supplied. They came to Bosch as perhaps the only remaining major leader of German industry who might still say what he really thought and asked him if he thought German industry was ready for war. He told the truth: He believed industry was not ready and that war at this point was impracticable. They asked him if he would be willing to say this to the highest levels of German government. According to at least one historian, he agreed. But when he tried to arrange a meeting with Göring, he was turned away.

That was what it had come to. Bosch was no longer relevant. The Nazi state had gotten what it wanted from him, and his opinion no longer mattered.

In May 1939 Bosch was asked to deliver a speech of welcome at the Deutsches Museum in Munich. That meant, of course, starting with a Nazi salute and the requisite words of tribute to Hitler. The night before the speech, he told a friend that he simply could not do it. When it was suggested that

Bosch fake an illness and let someone else speak in his place, he agreed. The next morning he appeared anyway. He was drunk. His friends could not stop him from taking his place at the podium. He launched into a slurred defense of the freedom and independence of science, and the importance of defending it from government interference. He mentioned Hitler only once, in what one attendee remembered as "a rather derogatory way." As he spoke, one Nazi after another in the audience rose, yelled at the speaker, and walked out.

Bosch was not arrested. It was not worth the headlines. Instead he was removed as chair of the museum's board and prohibited from further speech making. His remarks at the event were published, but only after they had been sanitized by Reich censors. He fell into a particularly severe depression, started suffering from physical symptoms, and spent time in a sanatorium. It seemed to revive him. Just as he was beginning to seem more his old self, Hitler invaded Poland. His bombs were made with Farben-generated explosives. His tanks and planes ran on Farben-made gasoline. After the invasion, Bosch fell completely from public view.

HIS ONLY FORAYS out of the house now days spent at a country retreat he maintained in the Black Forest, or an occasional motorcar ride into the countryside around Heidelberg. Bosch sat by himself in the back seat of his massive Maybach limousine, the chauffeur his only companion. He still enjoyed collecting plants and insects, and stopping at simple country inns for a meal. He sometimes left extravagant tips. He also enjoyed visiting the zoo: He had helped plan and in great part funded the Heidelberg Zoo, where he now spent hours staring at the animals, studying their behavior.

His physical health again declined, and his wife Else became worried about his mental health as well. He was at a lakeside retreat trying to regain his strength when, around Christmas 1939, his worried family called a doctor. Bosch had sequestered himself in a room, they said, alone, with some guns, and refused to let anyone in. The doctor managed to talk him out, and his family got him back to the villa in Heidelberg.

"It's a terrible gift when one can foresee the future," he told his family. "I have it, and what I see is horrible. My entire life's work will be destroyed, and I cannot survive that." He gave his wife an expensive piece of jewelry and told her, "Who knows if this won't be the last you will have to live on some day?" Before Hitler's war was over, he said, "You'll be glad if you'll be able to at least warm your hands on a teakettle."

In February 1940 Bosch suddenly moved to Sicily, leaving Else in Heidelberg, taking with him only some clothes and an ant colony that had been given to him by the Kaiser Wilhelm Institutes. He found no relief, began to complain about severe rheumatism, and was ordered home by his doctors. When he returned, he was diagnosed with pleurisy. He no longer cared. His only solace was alcohol. When his doctors told him to go to a sanatorium for better care, he told them no. "I've had enough," he said. "I don't want to go on."

He went into his office, gathered all his correspondence, and burned it.

In late April 1940, a few weeks after Hitler ordered the invasion of Denmark and Norway, Bosch called his son to his bedside. He had just enough energy to tell him what he foresaw. "To begin with, it will go well," he said. "France and perhaps even England will be occupied. But then he will bring the greatest calamity by attacking Russia. Even that will go well for a while. But then I see something horrific. Everything will be totally black. The sky is full of airplanes. They will destroy the whole of Germany, its cities, its factories, and also the IG."

Toward the end, Bosch said, "The clock's run down! It's over!" He died on April 26.

• • •

IT HAPPENED AS Bosch predicted: On May 12, 1944, the sky above his machine-city blackened and filled with airplanes. The U.S. Eighth Air Force had sent 220 bombers and their fighter escorts to destroy Leuna. Bosch's dream factory was now three square miles in size and thirty-five thousand workers strong. More than a quarter of the workers were prisoners and slave laborers. The plant was making one-quarter of all the Nazis' gasoline, including most the high-test fuel for Hitler's Luftwaffe, as well as fixed nitrogen for explosives and fertilizer, and artificial rubber for staff cars and troop transports. It was the single most important strategic target in Germany and the most heavily defended piece of land in Europe. It had blast walls and camouflage smoke, dummy plants and fields full of German fighter planes. The management trained five thousand Leuna workers as firefighters and twenty thousand more to operate antiaircraft guns and smoke pots. In most places in Germany, an antiaircraft battery consisted of six to a dozen guns. Leuna had *Grossbatteries*, superbatteries of thirty-two radar-controlled guns each, designed to work together, then trained the batteries not to shoot at individual planes but to fill big chunks of the sky with explosions, to create boxes of flak, zones of shrapnel so thick that no bomber flying into it could survive. Leuna had better air defenses than Berlin, and, as one American flyer said after a Leuna attack, "Boy, could they shoot."

When enemy bombers approached, the Germans unleashed a storm of metal. The sky turned dark with smoke, so black in places that the Americans could not see their targets and sometimes dropped their payloads miles from the factory. Planes limped back to England peppered with hundreds of holes—if they came back at all. During one string of Allied raids, 119 planes were lost without a single bomb falling on Leuna. U.S. flyers hated flying there. They nicknamed it "Flak hell Leuna," and the nearby town of Merseburg "Miseryburg." They had a greater chance of getting shot down over Leuna than anywhere else in the war. But they were ordered to attack again and again through the summer and fall of 1944, because it was becoming clear that Leuna was a key to victory.

In the hands of the Nazis, the synthetic gasoline project had succeeded beyond Bosch's dreams. In 1939, as Hitler's forces started their march across Europe, Germany was importing more than two-thirds of its fuel. Now, thanks to an expanded Leuna and the construction of more high-pressure synthetic fuel plants, Germany was making nearly three-quarters of its own gasoline. In the early months of 1944, after more than four years of continuous warfare, Germany's fuel stocks were as high as they had been since 1940. No nation since has matched the output of synthetic gasoline the Germans achieved in 1943.

The first huge Allied bomber attack in May 1944 was a surprising success, killing 126 workers and damaging enough critical machinery to temporarily shut down the plant. But a few days later reconnaissance photos showed the damaged areas being rebuilt. A second raid launched May 28 was less successful. Leuna was back to 75 percent of its original production level by July. The Nazis knew that their armed forces depended on gas from Leuna, so as the air attacks continued they gathered an army of 350,000 workers, including 7,000 engineers from the armed forces, all devoted to keeping the plant functioning. They dug better air-raid shelters, built better blast walls, and cannibalized needed parts from other plants. As fast as the Allies knocked out parts of Leuna, they fixed them. The German motto was "Everything for oil."

It was, as Reich minister of armaments and production Albert Speer put it, a race between concrete and bombs. What came to be called the Battle of Leuna was fought over the course of almost a year, with wave after wave of bombers hitting the area as long as the weather permitted. On July 7 they once again shut Leuna down. Production started again two days later. On July 19 a U.S. attack cut production by half. The next day, another attack shut the plant down for three days. A week after than, another attack cut production by two-thirds. Allied bombers were being shot down by the score, but the relentless attacks slowly began to grind Leuna down. Each raid took out a few more antiaircraft guns, downed a few more defending fighter planes, and destroyed a few more fire trucks. The vibrations from the explosions started loosening bolts in the plant, weakening walls, and springing leaks. Repairs became harder and harder.

By late October, the U.S. Eighth Air Force felt like it was winning. They were pounding Leuna into rubble. The flak was lightening, just a little, and the number of German defensive fighters was waning. For three weeks attacking bombers met hardly any enemy fighter planes over Leuna.

There was hope that another massive bomber attack on November 2 might finish it off. But the Germans had been hoarding aviation fuel, saving their planes, and waiting to spring a trap. Swarms of German fighters rose to meet the Americans, more than four hundred German planes, including some of the new "rocket planes"—early jets—that would fly almost six hundred miles per hour and ran rings around the U.S. fighters. Nearly seven hundred B-17s made that flight to Leuna. Fewer than four hundred made it back to base. Most of the rest were damaged.

It was a last gasp. The rocket planes used so much fuel that they could fly for only about eight minutes on a tank—and now they could no longer fill their tanks. As Leuna (and the other German synthetic gasoline factories targeted by the Allies) slowly began failing, Germany began starving for fuel. The Luftwaffe was especially hard hit. Pilot training was cut back because there was no fuel. Speer heard reports of air fields where trained flyers had only enough fuel to get into the air every third day. The army began suffer as well. Speer told his government of a column of 150 German military trucks in Italy being pulled by oxen.

After the war Speer testified that if the Allies had done nothing but destroy Leuna and the other synthetic fuel plants by bombing them day and night, the war would have been over in eight weeks.

Before it was over, the Allies launched twenty-two massive air raids on Leuna, in which more than six thousand bombers dropped more than eighteen thousand tons of explosives—an amount equivalent to the Hiroshima atomic bomb. They never succeeded in shutting Leuna down entirely—when the war ended, the plant was producing synthetic gasoline at about 15 percent of its former level—but they did cut off enough of Hitler's fuel supplies to hasten the end of the war. Bosch's dream machine was finally smashed and with it the hopes of the Third Reich.

But the machine would rise again and spread around the world, plants more productive than Leuna and more efficient than Bosch could imagine. Freed from military use, his high-pressure technology would be used again for what Bosch had originally intended: to feed the people of the world.

CHAPTER 22

THE MASS STARVATION that Crookes had predicted finally began to hit, it seemed, in 1958, when millions of people in China started to starve. Their revolutionary communist leader Mao Zedong was instituting the Great Leap Forward, a grand scheme to remake Chinese society by, among other things, uprooting traditional habits and lifestyles, moving millions of people from farm to city to work in factories, outlawing private garden plots, and creating enormous agricultural communes—reforming agriculture to conform to Maoist politics. His timing was inauspicious. The Great Leap Forward was launched just when drought, floods, and enormous population growth combined to tighten belts around China. When, on top of it all, Mao started enforcing his new policies, the agricultural system in some areas collapsed. Starvation began to spread. Hungry peasants began eating their livestock. When the animals were gone, they turned to wild vegetables, made grass soup, and in some places stripped the bark from trees and ate it. Cannibalism was reported. Despite everything, by 1961 an estimated thirty million Chinese had died from malnutrition. Instead of a great leap forward, Mao had triggered China's great famine, the worst mass starvation in recorded history.

China's leaders retrenched, stepping back to more traditional, more productive farming methods, and managing for the most part to recover. Ten years later, however, it looked like another disaster was coming. No matter how much food Chinese farms produced, it disappeared into the mouths of the plant's fastest-growing population. It was impossible for farmers to keep up with population growth. By the early 1970s most Chinese were relegated to eating a maintenance diet of rice plus a few vegetables. Meat was becoming a luxury. Food rationing was practiced in major cities. The average diet in China in 1970 was worse than it had been a generation earlier. All it would take was one big flood, one long drought, to tip the world's most populous nation into another mass famine.

That was thirty-five years ago. Today, China is battling a significant increase in obesity. The difference is Haber-Bosch.

The first major commercial transaction the Chinese government made after Richard Nixon's historic 1972 visit to Beijing was to order thirteen of the world's biggest, most modern Haber-Bosch fixed-nitrogen plants. It was the biggest business deal the Chinese had made with the West since the communists came to power. The machinery was delivered and built, the Chinese were trained to operate it, and within a few years the amount of fertilizer available in China more than doubled. Agricultural production shot up. More fertilizer plants were built.

Today China is both the world's single largest producer and the world's largest consumer of synthetic fertilizer. The population is much larger than it was in Mao's day (it has, since the time of the famine, added about as many people as there are in the United States, plus Mexico), but they are eating much better than they did a generation ago. Despite the increase in population, both average calories-per-person and the variety of available foods have both increased significantly. Instead of mass famine, the Chinese are now worried about overweight children.

TODAY HUNDREDS OF huge Haber-Bosch plants are drinking in air and turning out ammonia, producing enough fertilizer not only to support a burgeoning human population but to improve average diets worldwide. All the plants run on the same principles Haber and Bosch pioneered and are filled with the same basic catalyst that Alwin Mittasch found almost a century ago. They are, however, ever larger and more efficient. In Carl Bosch's day, the tallest ammonia ovens were thirty feet high. Now they top one hundred feet. In 1938, it took an average of sixteen hundred workers to produce a thousand tons a day of ammonia. Today it takes fifty-five workers to make the same amount. In the early days it took four times as much energy to make a ton of fertilizer as it does now. Still, the demand for their products is so great that Haber-Bosch plants today consume 1 percent of all the energy on earth, and the largest factories produce so much ammonia that it has to be transported in pipelines (one of the first ammonia pipelines in the United States, built in the late 1960s, for instance, runs from the plant in Texas to the corn fields of Iowa). This huge, almost invisible industry is feeding the world. Without these plants, somewhere between two billion and three billion people—about 40 percent of the world's population—would starve to death.

There are still pockets of starvation, to be sure; people die from hunger by the thousands every year, almost always because of a combination of local crop failures (often tied to droughts, floods, or warfare) and the disruption of transportation systems (again, owing mainly to armed conflict). The problem is never a global lack of food. There is plenty of food in the world. The problem is getting it to the people who are hungry.

The number of humans on earth rose from about 1.6 billion in 1890, when Crookes was predicting the end of civilization, to more than six billion today. While the population nearly quadrupled during the twentieth century, food production—thanks first to Haber-Bosch, second to improved genetic strains of rice and wheat—increased more than sevenfold. That is the simple math behind today's era of plenty.

A more personal way of gauging the impact of Haber-Bosch

is to look at your own body. About half the nitrogen in you came out of a Haber-Bosch factory. Don't worry: Nitrogen is nitrogen, the atoms in Haber-Bosch ammonia are precisely the same as the atoms in the best natural manure, and they all come, one way or another, from the air you breathe—but half the nitrogen in your blood, your skin and hair, your proteins and DNA, is synthetic.

AVOIDING FAMINE IS the good news. The not-so-good news is that the flood of synthetic nitrogen from Haber-Bosch plants is ending up in more places than food—like in our air and water. As a Stanford ecologist puts it, "We can't make food without mobilizing a lot of nitrogen, and we can't mobilize a lot of nitrogen without spreading some around."

Before Haber-Bosch, there were only two ways to get nitrogen out of the air and into food. One was lightning. But the most important one was the slow, steady process by which a few types of bacteria ate atmospheric nitrogen, broke it apart, and reformed it into substances plants could eat. The process is called bacterial nitrogen fixation. Some of these bacteria set up homes in nodules attached to the roots of plants, notably legumes like peas and beans, forming a symbiotic relationship in which they exchanged their fixed nitrogen for sugars and other food provided by the plants. These bacteria, working for millions of years, slowly built a stockpile of fixed nitrogen that fed most of the earth's plants, which fed all the animals. Life on earth depended on that stock of fixed nitrogen.

Haber-Bosch turbocharged the process. Today, Haber-Bosch plants produce an amount of fixed nitrogen equivalent to that produced naturally, doubling the amount available on earth. While this massive change in natural cycles means little to the basic composition of the atmosphere—there is so much N_2 in the air that the amount used by Haber-Bosch is negligible—it does mean a great deal to the biosphere, the places on the earth where life dwells.

Think of nitrogen atoms as riding on a series of big circles, like fairgoers on Ferris wheels. The wheel representing the natural, pre–Haber-Bosch "nitrogen cycle" begins in the air, with N_2 molecules. The next step is to break apart the N_2, fix the nitrogen, and begin moving it through living systems, starting with either bacteria (the process of biological nitrogen fixation) or with lightning strikes ripping apart the atmospheric nitrogen and combining it with other elements. Once it is fixed and available for living things, the nitrogen is passed from molecule to molecule and organism to organism, from bacteria to plants and plants to animals, in a series of subcycles, wheels within wheels. The living organisms release the fixed nitrogen back into the dirt when they die and rot. Some of it goes back into plants and cycles again, and some of it returns to the air (different types of bacteria can reverse the process, turning fixed nitrogen back into N_2). The intricacy and interconnectedness of these cycles, the paths spreading from the air to the land to the water and back, from nonliving to living systems, makes tracking difficult, research costly, and predictions almost impossible.

We know only that Haber-Bosch has altered these cycles enormously by injecting the world with a gigantic dose of synthetic nitrogen. It is as if we made our planet the subject of an experiment, doubling its food to see what would happen. Scientists are just beginning to grapple with the results.

Certain effects are easier to track than others. Researchers estimate, for example, that about half of all Haber-Bosch fixed nitrogen ends up in our air and water, not in our food. Say that a farmer dumps a ton of synthetic fertilizer on a field. Half of it feeds the crops, some of it goes back into the air, and most of the rest dissolves in the rain or irrigation water, leaches into the ground, and ends up in streams and lakes. Again, the nitrogen is mobile and can be incorporated into many different molecules in different ways—but much of it enters water systems in the form of various nitrates.

The level of nitrates in the Mississippi today is four times what it was in 1900. Levels in the Rhine are double those in the Mississippi. It does not all come from farmers; there is manure runoff from ranches (the manure rich in nitrogen from animals grown on feed made by crops fed with Haber-Bosch fertilizer) and municipal wastewater with its load of excess lawn fertilizers and home sewage. While nitrogen in water is not highly toxic in the sense of poisoning people, it can rise to dangerous levels. Nitrate pollution in water has been linked to health problems like methemoglobinemia, or "blue baby" syndrome. The precise role of high nitrate levels in disease is still, however, uncertain.

What is known is that nitrogen pollution in the water ends up feeding blooms of algae and weeds that turn waterways green and cloudy. It can get so bad that it cuts off sun reaching the depths, killing life below. As the vegetation dies and rots, it pulls oxygen out of the water. As oxygen levels go down, bottom-dwelling animals, shellfish and mollusks, begin to die off. The animals that feed on them starve. Toxins begin to collect. Freshwater systems begin to die.

Then the nitrates hit the ocean. Somewhere around 1.5 million tons of fixed nitrogen flows into the Baltic Sea north of Germany every year, making it one of the most polluted marine systems on earth. Oxygen levels are so low in some areas of the Baltic that the bottom is blanketed with mats of oxygen-hating bacteria. The Baltic cod industry collapsed in the 1990s. The same thing is beginning to happen worldwide. The Great Barrier Reef in Australia is starting to show the effects of vastly increased fertilizer use, and so is the Mediterranean, and so is the Black Sea.

But the biggest and best known of all the nitrate-affected waters in the world is the Dead Zone off the coast of Louisiana in the United States, where the Mississippi and Atchafalaya rivers flow into the Gulf of Mexico. Those two river systems drain more than 40 percent of the continental United States, including some of the most intensively fertilized farmland in thirty-one states. Nitrate levels in the gulf have more than doubled in the past forty years. At the same time, life in the water has increasingly begun to suffocate. The Dead Zone is a region where aquatic plants bloom, clams and lobsters die, fish flee, and the entire ecology has changed. The zone in the gulf fluctuates in size with the seasons, but it is now, on average, roughly the size of New Jersey. Every year it gets a little bigger. More than 150 smaller dead zones have been identified around the world, from Chesapeake Bay to the coast of Japan. Nitrogen pollution is not the only cause—water currents, temperatures, and natural fluxes in the growth of plants and animals also play a part—but the role Haber-Bosch plays is critical. Here again, much remains to be learned.

Just as poorly understood are the effects of Haber-Bosch nitrogen in the air. Some of it ends up in the air not as innocuous N_2, but as nitrogen-containing pollutants, including the notoriously dirty nitrogen oxides. Most of the oxides polluting our air come from burning fossil fuels in car engines and factories. But a portion, somewhere between 15 and 50 percent depending on how you count it and how far down the food chain you go, come directly or indirectly from Haber-Bosch plants. Spread fertilizer on a field, and a portion of it goes directly into the air, much of it from bacteria that break down the fixed nitrogen and release it as gases. This is especially true in flooded fields, like rice paddies. Again, some of the gas is innocuous N_2, the same atmospheric gas that started the process, a completion of a cycle. But part of it is released into the air as nitrous oxide, for instance, a potent greenhouse gas. Again, we have much to learn. No one knows exactly how much fixed nitrogen from Haber-Bosch products ends up in the air, or how much ends up in exactly what molecular forms. All we know is that it is a lot.

Of course, the pollutants don't just stay there. They fall or get washed back to earth in the rain, and when they do fall, they become fertilizers. What Haber-Bosch has done (along with the nitrogen oxides released by burning oil and coal) is to turn our atmosphere into a huge fertilizer silo. We are the beneficiaries of a bounty that the ancients could only dream about: tons of growth-promoting fertilizers showering from the sky. The amount of fixed nitrogen filtering down to earth in some places has risen so high that it equals the amount American farmers typically apply to their spring wheat. Again, the long-term effects of this worldwide change are uncertain. At first blush, it might seem like a great thing, the air enriched, a simple way to green the earth.

But the news is not all good. Nitrogen oxides (along with sulfur compounds in the air) create acid rain. That problem has received a lot of attention. Less attention has been paid to the issue of airborne fertilizer altering the amount of fixed nitrogen available to every ecosystem on the globe, from tundra to jungle, forest to prairie, ocean to desert. Some early studies indicate that ecosystems fed with the extra nitrogen begin, as expected, by flourishing. Then they reach a sort of saturation point in which at least some forests move into a "destabilization phase" in which productivity falls. If that happens, the system is less able to consume nitrogen, and more of what falls goes into the water. Soil chemistry changes. Species distributions change as nitrogen-hungry varieties outmuscle those naturally adapted to lower nitrogen levels. Natural systems are thrown awry.

Finally, questions remain about how the flood of Haber-Bosch nitrogen is affecting human agriculture and global populations. The easy availability of synthetic fertilizer has resulted in an increase in massive-scale monoculture farming, with the additional corn and other grains making possible huge animal factories. Haber-Bosch is not the only reason for our current agricultural system—increasing mechanization and advances in plant genetics are important as well—but it plays a major role. Most humans have moved past the old traditional methods of crop rotation and manuring, severed the old ties between crops and domestic animals, increased average farm sizes and decreased crop varieties. These developments offer terrific benefits, feeding many more people much better than our ancestors thought possible—but they also carry clear risks for soil quality, plant disease, and decreased diversity.

EVERY SCIENTIFIC DISCOVERY is two-edged. We have been so successful at meeting Sir William Crookes's challenge that humans now threaten to overrun the earth, just as nitrogen-eating algae and water plants overrun ponds and seas. Thanks to our recent access to endless fertilizer and the development of hardy, high-yield varieties of rice and wheat (the so-called Green Revolution of the late twentieth century) we are confronted with an abundance of food unprecedented in human history. Thomas Malthus has been confounded. Ours is an age marked not by mass starvation but by the easy availability of cheap, high-calorie food. Our health problem is not malnutrition, but conditions related to overweight, from diabetes to heart disease. We are dying of plenty. And the problem is not limited to the most developed nations. One-quarter of all adults in Thailand are overweight, and about a third of the residents of Beijing, and half the men and more than half the women in Mexico. This too is the legacy of Haber and Bosch.

WE NEED TO know much more. We need to learn about the uses and misuses of synthetic fertilizers, about the best ways to apply them to avoid pollution, about the care of soils and the changes in ecosystems. We need to understand the effects of our tinkering with the great machine of the world.

There are great mysteries here, and the promise of great

benefits. The earth's complex nitrogen cycle, the flow of changing forms as the element transitions out of the air and into living systems, gas to solid, solid to gas, inorganic to organic and back again, and the many organisms and environments that play a role, is only one of the great elemental cycles. Nitrogen interacts in as yet unknown ways with the carbon cycle, also complex, and crucially involved in global warning. As plants grow and die, for instance, they take up and release not just nitrogen but carbon. These linked cycles mean that the nitrogen produced in Haber-Bosch plants has an effect on global warming—although we don't know how much. The cycles dip in and out of the living world, in and out of the soils and waters, up to the sun's heat and down into the sunless depths of the sea, twisting on themselves and interacting with other. They turn and pulse and shift like living things, and their intricate webs feed all living things, including all of us. We need to study these great, complicated systems, appreciate their importance, and make sure our activities do not threaten them.

It is a funny thing: When you look at charts of these elemental cycles they begin to look like diagrams of the feedback systems in complex machines, or flow charts of a project through a huge business organization. They begin to look like challenges for the next Fritz Haber, the next Carl Bosch.

EPILOGUE

THE ATACAMA DESERT is again silent. After a last attempt by the Guggenheim family to revive the Chilean nitrate industry in the 1920s ("riches beyond the dreams of avarice" one enthusiastic Guggenheim wrote), the mills lost their race with Haber-Bosch and began to die, victims of the Depression and a world that no longer needed the region's *salitre*. For decades the abandoned mills were visited only by vandals, who rifled through the buildings, scrawled messages on the walls, and dug up the cemeteries. Two of the larger *oficinas* recently were declared international historical sites. A few tour buses labor their way up the steep hills from Iquique every week to show them to tourists. The Atacama remains rich with nitrates. While the richest of the original deposits have been worked through, the mills left enormous reserves of lesser-quality nitrate in the ground. Thanks to Haber-Bosch, however, no one is inclined to mine them. Iquique today, still featuring its turn-of-the-century main square, opera house, and old wood-front buildings, looks in places as if it was frozen in time in 1920. Along the waterfront, however, there are new hotels and casinos, discos and restaurants. Iquique is trying to remake itself into a resort town.

And what of the men whose lives were changed by Haber-Bosch? Kaiser Wilhelm II lived in Holland for more than two decades after his abdication. The former king bought a large villa, furnished it with trainloads of possessions from Germany, grew a beard, chopped wood for exercise, spent much of his time hunting, and entertained well-known and powerful visitors. When he grew bored, he sketched plans for grand buildings and unbuildable battleships. His conversations continued to weave insight and megalomaniacal diatribe. When the Nazis captured Paris, Wilhelm sent Hitler a congratulatory telegram. He died in 1941.

Carl Bosch's son shipped his precious collections—Bosch's insects, rocks, and gems, the treasures with which he filled thirty-eight oak cases in his villa in Heidelberg—to New York after the war. The twenty-five-thousand-specimen Bosch mineral collection was purchased by the Smithsonian in Washington, D.C. One suitcase of gems disappeared somewhere between Germany and the United States, and remains unaccounted for. Bosch left a final little mystery with his minerals: Attached to each specimen was a small label on which he penned a minor bit of code—quickly found to be a letter sequence tied to the word *amblygonit* (in English: amblygonite, a mineral) in which each letter in the word corresponds to a number from 1 to 9, with the final *t* standing for 0. When translated, the code yields nothing more important than the amount Bosch paid for each sample ("mtt," for instance, is 200 marks). Why he felt it important to hide the information with a code is unguessable.

Bosch's collection of more than four million insects also ended up at the Smithsonian. His private telescope, solace during so many long nights, went to the University of Tübingen, where it is still in use. His large private science library was shipped from Heidelberg and warehoused in Brooklyn after the war. Then someone neglected to pay the storage fee and the books ended up in the hands of a secondhand dealer, who broke up the collection and sold the books piecemeal.

Fritz Haber's research into chemical insecticides led to the an effective way to delouse buildings, which was further developed by other researchers into a poison gas called Zyklon B. It was used by the Nazis to kill concentration camp inmates. Farben held more than 40 percent of the stock of the company that made the gas.

"IG Farben was Hitler and Hitler was IG Farben," an American senator said after the war. In addition to producing fuel and explosives as outlined in this book, Farben's Buna artificial rubber program (greatly speeded and expanded after Bosch's death) proved vitally important to Hitler's plans. In the final months of the war Farben was feverishly trying to finish a giant Buna plant next to Auschwitz, a futile effort during which thousands of Jews, political prisoners, and foreign slave laborers were literally worked to death. Twenty-three Farben executives were tried for war crimes at Nuremberg on charges that included conspiracy in the planning and execution of the war and enslavement and murder of civilian populations. Carl Krauch, Hermann Schmitz (who took over Bosch's position), and eleven other company executives were given prison sentences ranging from eighteen months to eight years. Ten other Farben executives were acquitted and the giant German cartel was broken into its constituent parts. A revived BASF appointed Else Bosch, Carl's widow, a member of the supervisory board in the 1950s. BASF has prospered in the postwar world, growing again into the world's largest chemical company (ranked by total sales).

Leuna was repaired and operated by the East Germans after the war. Today it is advertised as Europe's largest chemical factory. Marketed as a cutting-edge "chemical park," it houses two dozen firms making everything from ammonia and synthetic fuels to plastics and pure gases—many of the same products pioneered by Haber and Bosch.

Haber's institute at Dahlem was renamed in his honor, and the Haber Institute is today an important scientific center in Germany. The Kaiser Wilhelm Institutes were renamed for Max Planck.

Hermann Haber, Fritz and Clara's son, committed suicide after the war.

Source Notes

A CHEMICAL CAST OF CHARACTERS

Atmospheric Nitrogen = N_2

The "Fixed" Nitrogens:
Ammonia = NH_3
Ammonium sulfate = $(NH_4)_2SO_4$
Calcium cyanamide = $CaCN_2$
Iron nitride = Fe_2N

The Nitrates:
Potassium nitrate = "true" saltpeter = China snow = KNO_3
Sodium nitrate = Chilean saltpeter = *salitre* = white salt = $NaNO_3$
Hydrogen nitrate = nitric acid = HNO_3
Calcium nitrate = Norwegian saltpeter = $Ca(NO_3)_2$
Ammonium nitrate = NH_4NO_3

Instead of extensive footnotes I offer the chapter-by-chapter summaries below, which list the most important sources as well as some additional details. To keep the book to a reasonable length (and in deference to more general, rather than scientifically inclined, readers) I have chosen not to include a great deal of detailed chemical information and much of the nitrogen history that preceded Haber and Bosch—

including almost all the early work on nitre, azote, mephitic air, and phlogiston, as well as the theories of the relationships among fire, air, and life—devised by dozens of brilliant researchers, from Paracelsus to Galileo and Torricelli, Boyle and Hooke, Lavoisier, Berthellot, Rutherford, Thomson, Priestley, Cavendish, Sheele, Davy, van Helmont, von Liebig, Lawes and Gilbert, and many others.

A few general points: primary information for this book was gathered from archives and museums in Germany and South America. Secondary materials include contemporary news stories, scholarly articles, and books cited in the bibliography. This volume offers biographical sketches of the two main characters, Fritz Haber and Carl Bosch, not full portraits. More complete treatments of Haber's life are available to readers in three English-language biographies (Charles 2005, Goran 1967, and an English translation of the German work by Stoltzenberg 2004)—and additional biographical information on Haber can be found in books by his ex-wife Charlotte, his son Ludwig Fritz Haber, and numerous other sources. Very little has been written about Carl Bosch. The only full-length treatment of Bosch's life is an admiring and somewhat selective German-language biography written more than a half century ago by a former associate (Holdermann 1954). I relied where I could on Bosch's own words, found in various German journals and papers, contemporary news reports, BASF archival holdings, and materials in the small but informative Bosch Museum in Heidelberg. His is a great, tragic life, important to the history of twentieth-century science and technology. I hope that the information in this book, which constitutes perhaps the most complete picture of Bosch available in English, will spur further interest.

I draw particular attention to Smil (2001), which is a wonderful book and one on which I depended throughout this project.

The abbreviated entries in the notes refer to full citations in the bibliography. In addition, I use the following acronyms:

BASF Archive, Ludwigshafen (BASFA)

Bosch Museum, Heidelberg (BM)

Museo Naval de Iquique (MNI)

CHAPTER I

Crookes had been interested in food and fertilizer since doing a bit of contract work for the City of London, in which he had been asked to find a way of using the city's rising tide of sewage to help farmers grow crops. He found no easy answers but started thinking about food production and nitrogen. At one point he also did experiments in which he fixed nitrogen by burning it out of the air. His dire prediction of mass starvation echoed that made more than a century earlier by the Reverend Thomas Robert Malthus, a British cleric who wrote about population outstripping food. Malthusian doom had been avoided by opening of vast new wheat-growing regions in the United States, Russia, and Canada, which grew enough grain to match population growth. By Crookes's time, Malthus was for the most part forgotten. Crookes spent about half his speech at Bristol talking not about food but about another controversial topic—the scientific study of life after death (he was at the time also the president of the Society for Psychical Research). Details of Crookes's speech on the wheat problem were drawn from his own words (Crookes 1898, 1900), and contemporary reports in the Bristol newspapers *Mercury* (Sept. 8, 1898, p. 5), *Western Daily Press* (Sept. 6, 1898, p. 7), and *Times-Mirror* (Sept. 9, 1898, p. 5). More information on Crookes and the effect of his challenge to chemists can be found in D'Albe (1924), Charles (2005), Fisher and Fisher (2001), Haber (1971), Leigh (2004b), Smil (2001), Travis (1998), and Waeser (1926). The paragraphs about premodern farming techniques were drawn primarily from Leigh (2004b) and Smil (2001).

CHAPTER 2

The early history of gunpowder and its relationship to saltpeter is detailed in Bown (2005), Kelly (2004), and Partington (1960), with additional information from Chilton (1968), Smil (2001), Waeser (1926), and Wisniak (2000). The legend of the devil's appearance in the desert was told to me by a naturalist in Iquique, Guillermo Rivero; it can also be found in Bown (2005) and Wisniak and Garces (2001). Darwin's visit to Iquique is drawn from his own words in his notes from the *Beagle* (Darwin 1987). General points about the Atacama and the early nitrate history of the Tarapacá are taken from my visits to the area, with additional information from Chilton (1968), Farcau (2000), O'Brien (1982), Smil (2001), and Wisniak and Garces (2001). Additional detail on the early nitrate trade is in Leigh (2004b).

CHAPTER 3

Descriptions of the Chinchas (especially the smell!) come from a visit I made to the area, as well as from contemporaneous accounts of the Chinchas during their heyday, found in Duffield (1877), Markham (1862), Peck (1854), Tschudi (1849), and Twain (1913), as well as U.S. government documents dealing with guano policy and the coolie trade. Other important sources are Blakemore (1974), Clayton (1999), Farcau (2000), Irick (1982), Leigh (2004b), Lubbock (1955, 1966), Masterson (2004), Mathew (1981), McCreery (2000), O'Brien (1996), Schwendinger (1988), Skaggs (1994), Smil (2001), Stewart (1951), and Wilmott (2004).

CHAPTER 4

There is still not absolute scientific agreement on how the nitrates built up in the Atacama. Early geologists followed Darwin's lead, supposing that an enormous inland sea evaporated and deposited the minerals, with the mix enriched perhaps as the shrinking sea became lined with millions of pounds of dead fish and rotting vegetation, which might have attracted millions of birds that added their dung, all of which dried and condensed into the mineralized crust. Or perhaps the spume of the Pacific was carried over the hills on the wind and slowly deposited its salts, microgram by microgram, for millennia. Others believed the minerals came from the Andes, washed down in salty, dead-end rivers. The most current theory focuses on slow, almost imperceptible accretion that might have happened in many places but resulted in big deposits in the Atacama thanks to the combination of aridity and frequent fogs. Some of the history of the growing nitrate trade can be found in Blakemore (1974) and Farcau (2000) and especially in Chilton (1968), Duffield (1877), Haynes (1954), James (1993), Leigh (2004b), Lubbock (1966), Mathew (1981), McCreery (2000), O'Brien (1982), and Wisniak and Garces (2001). A visit to Iquique helped me understand the people and their history. Farcau (2000) is a history of the War of the Pacific; most of the sources listed earlier for this chapter contain additional information. In addition, see Burgess and Harbison (1954), Clayton (1999), Klein (2003), and Sater (1973). I became fascinated with the American adventurer and entertainer Paul Boyton, who, in addition to his exploits in Peru, attracted huge crowds by swimming seemingly impossible stretches of water (across the English Channel, down the Mississippi) wearing an inflatable rubber suit with a little sail attached to his foot. Boyton later started an amusement park that became

Coney Island. Readers interested in learning more about him can read Lyon (1960) and Sobol (1975).

CHAPTER 5

The growth of the Chilean nitrate industry after the War of the Pacific is discussed in Farcau (2000), Leigh (2004b), McCreery (2000), O'Brien (1996), Smil (2001), and Wisniak and Garces (2001). Information on that era and especially John Thomas North, "the Nitrate King," can be found in Blakemore (1974), O'Brien (1982), and Monteon (1975). Bown (2005) provides a good overview of the development of high explosives. Allen (1978) describes the nitrate clippers, as does Lubbock (1966). Information on the *ficha* system and the Iquique massacre comes from museums in the area, as well as Barrios (1998), Collier and Sater (2004), and Deves (1997).

CHAPTER 6

I relied on Charles (2005), Goran (1967), and Stoltzenberg (2004) for many biographical details of Haber's life. Haber's friend Richard Willstätter also included a great deal of good information on Haber in his memoir (Willstätter 1965). Other sources on Haber's early life and chemistry include Hoffmann (1995), Perutz (1998), and Smil (2001); Elon (2002) and Stern (1999) are especially good on the Jewish experience in Wilhelmine Germany. Appl (1982), Haber (1971), Leigh (2004), A. S. Travis (1998), and T. Travis (1993) provide additional information on the state of chemistry, the controversy with Nernst, and the first steps toward the ammonia synthesis. Barkan (1999) is Nernst's biographer.

CHAPTER 7

Information on Wilhelm Ostwald came from Johnson (1990) and Smil (2001), as well as the Haber biographies. A detailed history of BASF, including the early nitrogen work, is provided by Abelshauser, Hippel, Johnson, and Stokes (2004). Additional information, including details on Brunck's career, can be found in BASF (2005), BASFA, BM, Borkin (1978), Haber (1971), Hayes (1987), Saftien (1958), and Stern (1999).

CHAPTER 8

Much of this chapter is based on the Haber biographies and general sources mentioned at the beginning of this section. Holdermann (1954) describes Bosch's introduction to the field and his run-in with Ostwald, and offers details on his early life. I relied on materials from BM, and other biographical sources for Bosch can be found at the beginning of this section. Bosch's first years at BASF and the early Haber-Bosch work are described in these basic biographical sources as well as in BASFA, Hayes (1987), Mittasch (1932), Nagel (1958), A. S. Travis (1998), and T. Travis (1993). The critical 1909 demonstration has been described in many places; I relied on the accounts by Charles (2005), Smil (2001), and Stoltzenberg (2004).

CHAPTER 9

In addition to the standard Haber-Bosch materials, information came from Abelshauser, Hippel, Johnson, and Stokes (2004); BASFA materials; Bosch (1932); Ernst (1928); Furter (1982); Nagel (1958); Perutz (1998); Stern (1999); and Stranges (1984). Chorkendorff and Niemantsverdriet (2003); Ertl, Knözinger, and Weitkamp (1999); Hughes (1969); Leigh (2004a and b), Mittasch (1951), and A. S. Travis (1998) provided additional material on the search for catalysts.

CHAPTER 10

I relied heavily on Bosch's own account of scaling up Haber's discovery (Bosch 1932). More information was gathered from his biography by Appl (1982), BASFA holdings under his and Haber's names, BM materials, Furter (1982), Haber (1971), Holdermann (1954), Leigh (2004a and b), Martin and Barbour (1915), Smil (2001), and T. Travis (1993). Haber's work immediately afterward is well described in Stoltzenberg (2004), as well as the other Haber biographies.

CHAPTER 11

The most complete description of the Hoechst lawsuit is found in Stoltzenberg (2004); I also used materials from Charles (2005) and Johnson (1990). Information about Haber's move to KWI came from the biographical materials, with additional material from Fuchs and Hoffman (2004), Haber (1986), and Willstätter (1965). BASF (2005) refers to the Oppau purchase, and more details of the construction of the plant are found in Abelshauser, Hippel, Johnson, and Stokes (2004); Bosch (1932); Furter (1982); Haber (1971); Leigh (2004b); Nagel (1958); Smil (2001);

and T. Travis (1993). Stories of Einstein's time at Dahlem and his relationship with Haber are from Elon (2002), Isaacson (2007), and Stern (1999), as well as the Haber biographies.

CHAPTER 12

Information about the importance of nitric acid as a raw material for explosives during World War I, the competition to provide it, and the ways in which BASF kept a dominant position—including the construction of Leuna—can be found in Abelshauser, Hippel, Johnson, and Stokes (2004); Chilton (1968); Ernst (1928); Haber (1971); the Haber biographies; Leigh (2004a and b); Meinzer (1998); Szollosi-Janze (2000); T. Travis (1993); and Waeser (1926). BASF (2005), Johnson (1990), and Smil (2001) provide additional background. The Spee naval battles are described in Bown (2005) and Haynes (1945). Spee's own report is available on the Web at http://net.lib.byu.edu/~rdh7/wwi/1914/spee.html. Details of the French air attack on Oppau in 1915 were gathered from contemporaneous reports from BASFA. Other information can be found in Abelshauser, Hippel, Johnson, and Stokes (2004). (The British also tried bombing raids in the Oppau area, later in the war.)

CHAPTER 13

My unflattering portrait of Kaiser Wilhelm is based on material by Roehl (1967, 1994) and Stern (1999). Additional context on Wilhelmine Germany was drawn from Elon (2002), Freese (1947), Leigh (2004b), Macmillan (2002), and Stern (2006). Haber's World War I work is detailed in his biographies, with additional material from Perutz (1998) and Willstätter (1965). Bown (2005) offers more on poison gas, as does Haber (1986), Johnson (1996), Lehrer (2000), and Trumpener (1975). I have followed the general path leading to Clara Immerwahr's suicide presented in the Haber biographies, supplemented by the thoughts of Perutz (1998) and Stern (1999). Ironically, the name Immerwahr can be loosely translated as "ever true."

CHAPTER 14

Abelshauser, Hippel, Johnson, and Stokes (2004) is the primary source for information on BASF during and just after the war. Other details come from Freese (1947), Haber (1971), Macmillan (2002), and Waeser

(1926). More information on the French occupation is taken from BASFA files, Borkin (1978), and Meinzer (1998). Supplementary material on Bosch's activities at Versailles can be found in Holdermann (1954) and Nagel (1958). The comical attempts of the British to uncover the workings of Oppau are found in T. Travis (1993) and Van Rooij (2005), and especially in Reader (1970).

CHAPTER 15

Information on Haber's second marriage and his fleeing to Switzerland is taken from Charles (2005) and Goran (1967). Stern (1999) calls Charlotte Nathan's memoir "entirely unreliable"; I avoided it. Haber's Nobel Prize speech is available online at http://nobelprize.org/nobel_prizes/chemistry/laureates/1918/haber-lecture.html. More on the Nobel comes from Bown (2005) and the Haber biographies; I used the biographies for information on Haber's search for gold as well, with additions from Hoffmann (1995) and Willstätter (1965).

CHAPTER 16

The development of Oppau and Leuna, including information on the Oppau explosion, is taken primarily from Abelshauser, Hippel, Johnson, and Stokes (2004), which also includes material on the worker takeover of Leuna. Additional information on the Oppau blast and its aftermath comes from BASFA documents, including Bosch's memorial speech (Bosch 1921), Hayes (1987), and Smil (2001). Haber (1971) and Lochner (1954) have more on Bosch's relations with his workers. The story of Bosch's breakdown is in Holdermann (1954).

CHAPTER 17

Borkin (1978), Lochner (1954), Meinzer (1998), Lefebure (1923), and Van Rooij (2005) offer details and context concerning the second French occupation of the BASF factories on the Rhine. The story of K and A, the industrial spies, is told in Reader (1970) and T. Travis (1993). Glaser-Schmidt (1994) offers more on the growth and dissemination of Haber-Bosch technology after World War I, as do Abelshauser, Hippel, Johnson, and Stokes (2004); Bosch (1932); Hayes (1987); Leigh (2004a and b); and Smil (2001), and Waeser (1926). The first moves toward synthetic gasoline are outlined in Hayes (2001) and Hughes (1969), as well as Abelshauser, Hippel, Johnson, and Stokes (2004); Borkin (1978); and Stranges (1984). Abelshauser, Hippel, Johnson, and Stokes (2004); Borkin (1978); and Hayes (2001) also review the formation of IG Farben. My section on Bosch's lifestyle in Heidelberg comes from a visit to his villa and BM, now installed in his former chauffeur's house, as well as and anecdotes from BASFA, Holdermann (1954), and Mittasch (1932).

CHAPTER 18

Haber material is drawn from his biographies as well as Johnson (1990), Perutz (1998), Stern (1999), and Stranges (1984); the Meitner quote is from Sime (1996). The journalist's view of Leuna is found in Crippen (1944). Abelshauser, Hippel, Johnson, and Stokes (2004) cover the changing financial situation at BASF in the 1920s, as do Hayes (2001) and Glaser-Schmidt (1994). Borkin (1978); Freese (1947); Haber (1971); Hughes (1969); Larson, Knowlton, and Popple (1971); and Reader (1970) provide perspectives on the maneuvering surrounding the synthetic gasoline project. The cozy relationships among Standard Oil, Ford, and IG Farben in the years before World War II is a story that has yet to be fully told.

CHAPTER 19

The sources for the synthetic fuels project noted in the previous chapter apply here as well. Bosch's politics are described in Hughes (1969) and mentioned in Hayes (1987), Lochner (1954), and Meinzer (1998). The general political situation and BASF's economic situation are found in Abelshauser, Hippel, Johnson, and Stokes (2004) and Hayes (2001). Haber's biographers outline his activities during the early days of the Depression; I also relied on Elon's (2002) and Stern's (1999) views of the rise of Hitler.

CHAPTER 20

Haber's response to Hitler's anti-Jewish edicts is found in the Haber biographies. I used additional material from Sime (1996) and especially Beyerchen (1977), Lehrer (2000), and Stern (1999). Hentschel (1996) offers a number of fascinating primary documents from the era. Bosch's and BASF's actions during the early days of Nazi control are taken from Abelshauser, Hippel, Johnson, and Stokes (2004); Borkin (1978); Freese (1947); Hallgarten (1952); Hayes (1987, 2001); Holdermann (1954); Hughes (1969); Lochner (1954); and Smil (2001).

Chapter 21

The Haber memorial is described in his biographies; I also used material from Beyerchen (1977), Heilbron (1986), and Holdermann (1954). The Holdermann book is a source of material for Bosch's final years at BASF, along with materials from BM and BASFA, Beyerchen (1977), Borkin (1978), Hayes (2001, 2003), Hughes (1969), and Lochner (1954). I drew my description of the Battle of Leuna during World War II from Grant (2007), McArthur (1990), D. L. Miller (2006), and U.S. Air Force (1987) reports.

Chapter 22

Smil (2001) provides the most important overview of the recent use and impact of Haber-Bosch technology, including information on the industry's growth and environmental impact. Additional information on the political and economic importance of synthetic fertilizer in China can be found in Becker (1996) and Charles (2005). On trends in fixed nitrogen use see Sahota (1968). For more information on the growing problem of nitrogen pollution, I looked to Fisher and Fisher (2001), Follett and Hatfield (2001), and Leigh (2004b).

Chapter 23

In 1936 about a third of the fixed nitrogen used by U.S. farmers came from Haber-Bosch plants. By 1960 it was 90 percent. The biggest Haber-Bosch plant in the world today, in Russia, produces about 2.5 million tons of ammonia every year—dwarfing the biggest plant in the United States, at Donaldsonville, Louisiana, which pumps out a scant 1.55 million tons. Information about the present-day impact of the discovery comes from Charles (2005), Fisher and Fisher (2001), Hoffmann (1995), Leigh (2004), Sahota (1968), and Smil (2001).

BIBLIOGRAPHY

Abelshauser, Werner, Wolfgang von Hippel, Jeffrey Allan Johnson, and Raymond G. Stokes. 2004. *German industry and global enterprise: BASF: The history of a company.* Cambridge, England: Cambridge University Press.

Aftalion, Fred. 1991. *A history of the international chemical industry.* Philadelphia: University of Pennsylvania Press.

Allen, Oliver E. 1978. *The windjammers.* Alexandria, VA: Time-Life Books.

Almqvist, Ebbe. 2003. *History of industrial gases.* New York: Kluwer Academic.

Appl, Max. 1982. The Haber-Bosch process and the development of chemical engineering. In *A century of chemical engineering,* ed. William F. Furter, 29–53. New York: Plenum Press.

Balfour-Paul, Jenny. 2000. *Indigo.* Chicago: Fitzroy-Dearborn.

Barkan, Diana Kormos. 1999. *Walther Nernst and the transition to modern physical science.* Cambridge, England: Cambridge University Press.

Barrios, Pablo Artaza, et al. 1998. *A 90 años de los sucesos de la escuela Santa Maria de Iquique.* Santiago, Chile: LOM Ediciones.

BASF Aktiengesellschaft Community Relations. 2005. *Historical milestones 1865–2005.* Ludwigshafen, Germany: BASF.

Becker, Jasper. 1996. *Hungry ghosts: Mao's secret famine.* New York: Free Press.

Beyerchen, Alan D. 1977. *Scientists under Hitler.* New Haven, CT: Yale University Press.

Blakemore, Harold. 1974. *British nitrates and Chilean politics: 1886–1896.*
London: Athlone Press.

Blottnitz, H. von, A. Rabl, D. Boiadjiev, T. Taylor, et al. 2006. Damage
costs of nitrogen fertilizer in Europe and their internalization.
Journal of Environmental Planning and Management 49: 413–33.

Borkin, Joseph. 1978. *The crime and punishment of I. G. Farben.* New
York: Free Press.

Bosch, Carl. 1921. Oppau memorial speech. *Werkzeitung der Badischen
Anilin-&Soda-Fabrik Ludwigshafen* 10 (October):139–40.

———. 1932. The development of the chemical high pressure method
during the establishment of the new ammonia industry [Nobel lec-
ture].

———. 1936. Why chemical industry is international. *Chemical and
Metallurgical Engineering* 43:250.

Bown, Stephen R. 2005. *A most damnable invention.* New York: St. Martin's Press.

Burgess, Eugene W., and Frederick H. Harbison. 1954. *Casa Grace in
Peru.* Washington, DC: National Planning Association.

Carpenter, F. B. 1909. The fixation of nitrogen. *Journal of Industrial and
Engineering Chemistry* 1:4–5.

Cerruti, F. E. 1864. *Peru and Spain.* London: Williams and Norgate.

———. 2005. *England's Leonardo.* Bristol, England: Institute of Physics
Publishing.

Chapman, Allen. 1996. England's Leonardo: Robert Hooke (1635–
1703) and the art of experiment in Restoration England. *Proceedings
of the Royal Institution of Great Britain* 67:239–75.

———. 2003. Restoration man. *Oxford Today* 15:1–4.

Charles, Daniel. 2002. The tragedy of Fritz Haber. NPR transcript,
July. www.npr.org/programs/morning/features/2002/jul/fritzhaber/

———. 2005. *Master mind: The rise and fall of Fritz Haber.* New York:
HarperCollins.

Chilton, Thomas H. 1968. *Strong water.* Cambridge, MA: MIT Press.

Chorkendorff, I., and J. W. Niemantsverdriet. 2003. *Concepts of modern
catalysis and kinetics.* Weinheim, Germany: Wiley-VCH.

Clayton, Lawrence A. 1999. *Peru and the United States: The condor and
the eagle.* Athens, GA: University of Georgia Press.

Collier, Simon, and William F. Sater. 2004. *A history of Chile: 1808–2002.*
Cambridge, England: Cambridge University Press.

Cornwell, John. 2003. *Hitler's scientists: Science, war, and the devil's pact.*
New York: Viking.

Crippen, Harlan. 1944. *Germany: A self-portrait: A collection of German writings from 1914 to 1943.* Oxford: Oxford University Press. (Includes "A Laborer in Leuna," a report from the newspaper *Berliner Tageblatt*, December 4, 1927, pp. 229–231)

Crookes, William. 1898. Address of the president before the British Association for the Advancement of Science, Bristol. *Science* n.s. 8:561–612.

———. 1900. *The wheat problem.* London: G. P. Putnam's Sons.

Crowell, Benedict, and Robert Forrest Wilson. 1921. *The armies of industry,* vol. 1, *Our nation's manufacture of munitions for a world in arms, 1917–1918.* New Haven, CT: Yale University Press.

Crowther, Samuel. 1933. *America self-contained.* New York: Doubleday, Doran and Co.

Dahl, Per F. 1999. *Heavy water and the wartime race for nuclear energy.* Bristol, England, and Philadelphia: Institute of Physics Publishing.

D'Albe, Fournier. 1924. *The life of Sir William Crookes.* New York: D. Appleton and Co.

Darrow, Floyd L. 1930. *The story of chemistry.* New York: Blue Ribbon Books.

Darwin, Charles. 1987. *Diary of the Voyage of HMS Beagle,* vol. 1. New York: New York University Press,

Davis, William Columbus. 1950. *The last conquistadores.* Athens: University of Georgia Press.

De Shazo, Peter. 1979. The Valparaiso maritime strike of 1903 and the development of the revolutionary labor movement in Chile. *Journal of Latin American Studies* 11:145–68.

Debus, Allen G. 1964. The Paracelsian aerial niter. *Isis* 55:43–61.

Deichmann, Ute. 2006. The kaiser's chemist. *Times Literary Supplement.* June 16.

Deves, Eduardo. 1997. *Los que van a morir te saluden.* 3rd ed. Santiago, Chile: LOM Ediciones.

DuBois, Josiah E., Jr. 1953. *Generals in grey suits.* London: Bodley Head.

Duffield, A. J. 1877. *Peru in the guano age.* London: Richard Bentley and Son.

———. 1881. *The prospects of Peru.* London: Newman and Co.

Elon, Amos. 2002. *The pity of it all.* New York: Metropolitan Books.

Elzen, Michel G. J. den, et al. 1997. The biogeochemical submodel: Cycles. In *Perspectives on global change: The TARGETS approach,* ed. Jan Rotmans and Bert de Vries, 161–88. Cambridge, England: Cambridge University Press.

Ernst, Frank A. 1928. *Fixation of atmospheric nitrogen.* New York: D. Van Nostrand Co.

Ertl, Gerhard, Helmut Knözinger, and Jens Weitkamp, eds. 1999. *Preparation of solid catalysts.* Weinheim, Germany: Wiley-VCH.

Farcau, Bruce W. 2000. *The ten cents war.* Westport, CT: Praeger.

Fischer, Conan. 2002. *The Ruhr crisis, 1923–1924.* Oxford, England: Oxford University Press.

Fisher, David E., and Marshall Jon Fisher. 2001. The nitrogen bomb. *Discover* 22:53–54.

Follett, R. F., and J. L. Hatfield. 2001. *Nitrogen in the environment: Sources, problems, and management.* Amsterdam: Elsevier.

Freese, Barbara. 1947. *IG Farben.* New York: Boni and Gaer.

———. 2003. *Coal: A human history.* New York: Penguin.

Fuchs, Eckhardt, and Dieter Hoffmann. 2004. Philanthropy and science in Wilhelmine Germany. In *Philanthropy, patronage, and civil society,* ed. Adam Thomas, 103–119. Bloomington: Indiana University Press.

Furter, William F. 1982. *A century of chemical engineering.* New York: Plenum Press.

Galloway, J. N., F. J. Dentener, D. G. Capone, E. W. Boyer et al. 2004. Nitrogen cycles: Past, present, and future. *Biogeochemistry* 70:153–226.

Gibbs, Antony and Sons. 1958. *Merchants and bankers.* London: Antony Gibbs and Sons Ltd.

Glaser-Schmidt, Elisabeth. 1994. Foreign trade strategies of I. G. Farben after World War I. *Business and Economic History* 23:201–11.

Gootenberg, Paul. 1993. *Imagining development.* Berkeley: University of California Press.

Goran, Morris. 1947. The present-day significance of Fritz Haber. *American Scientist* 35:400–3.

———. 1967. *The story of Fritz Haber.* Norman: University of Oklahoma Press.

Grant, Rebecca. 2007. Twenty missions in hell. *Air Force Magazine* (April):74–78.

Guerlac, Henry. 1954. The poets' nitre. *Isis* 45:243–55.

Haber, Fritz. 1914. Modern chemical industry. *Journal of Industrial and Engineering Chemistry* 6:325–31.

———. The synthesis of ammonia from its elements. [1920 Nobel lecture in chemistry]. 326–40.

Haber, L. F. 1971. *The chemical industry 1900–1930.* Oxford, England: Clarendon Press.

———. 1986. *The poisonous cloud: Chemical warfare in the First World War.* Oxford, England: Clarendon Press.

Hallgarten, George W. F. 1952. Adolf Hitler and German heavy industry, 1931–33. *Journal of Economic History* 12:222–46.

Harper's Weekly. 1865. Spain and Chili. December 9, pp. 780–81.

Hayes, Peter. 1987. Carl Bosch and Carl Krauch: Chemistry and the political economy of Germany, 1925–1945. *Journal of Economic History* 47:353–63.

———. 2001. *Industry and ideology: IG Farben in the Nazi era.* Cambridge, England: Cambridge University Press.

———. 2003. Profits and persecution: German big business and the Holocaust. J. B. and Maurice C. Shapiro Annual Lecture, February 17, 1998, United States Holocaust Museum Center for Advanced Holocaust Studies. Washington, DC: United States Holocaust Museum.

Haynes, Williams. 1945. *American chemical industry, the World War I periods: 1912–1922,* vol. 2. New York: D. Van Nostrand Co.

———. 1954. *American Chemical Industry, The World War I Periods: 1912–1922,* vol. 1. New York. D. Van Nostrand Co.

Heilbron, J. L. 1986. *The dilemma of an upright man: Max Planck as spokesman for German science.* Berkeley: University of California Press.

Heiserman, David L. 1992. *Exploring chemical elements and their compounds.* Blue Ridge Summit, PA: TAB Books.

Hempel, Edward H. 1939. *The economics of chemical industries.* New York: John Wiley and Sons.

Hentschel, Klaus, ed. 1996. *Physics and national socialism: An anthology of primary sources.* Basel, Switzerland: Birkhaeuser Verlag.

Herschback, Dudley. 1998. Teaching chemistry as a liberal art. *Harvard University Gazette.* www.hno.harvard.edu/gazette/1998/05.21/TeachingChemist.html.

Hoffmann, Roald. 1995. *The same and not the same.* New York: Columbia University Press.

Hoffmann, Roald, and Pierre Laszlo. 2001. Coping with Fritz Haber's somber literary shadow. *Angewandte Chemie International Edition* 40:4599–604.

Holdermann, Karl. 1954. *Im Banne der Chemie: Carl Bosch, Leben und Werk.* Dusseldorf: Econ-Verlag.

Hounshell, David A. 1988. *Science and corporate strategy: Du Pont R&D, 1902–1980.* Cambridge, England: Cambridge University Press.

Houston, Edwin J. 1894. *Electricity, one hundred years ago and today.* New York: W. J. Johnston.

Howard, Frank A. 1947. *Buna rubber: The birth of an industry.* New York: D. Van Nostrand Co.

Hughes, Thomas Parke. 1969. Technological momentum in history: Hydrogenation in Germany 1898–1933, *Past and Present* 44:106–32.

Irick, Robert L. 1982. *Ch'ing policy toward the coolie trade, 1847–1878.* China: Chinese Materials Center.

Isaacson, Walter. 2007. *Einstein: His life and universe.* New York: Simon & Schuster.

Jago, Lucy. 2001. *The northern lights.* New York: Alfred A. Knopf.

James, Marquis. 1993. *Merchant adventurer: The story of W. R. Grace.* Washington, DC: SR Books.

Jenkinson, D. S. 2001. The impact of humans on the nitrogen cycle, with focus on temperate arable agriculture. *Plant and Soil* 228:3–5.

Johnson, Jeffrey Allan. 1990. *The kaiser's chemists.* Chapel Hill: University of North Carolina Press.

———. 1996. The scientist behind poison gas: The tragedy of the Habers. *Humanities* (November/December): 25–29.

———. 2004. The power of synthesis (1900–1925). In *German industry and global enterprise: BASF: The history of a company,* ed. Werner Abelhauser, 115–205.

Johnson, Jeffrey Allan, and Roy MacLeod. 2007. The war the victors lost: The dilemmas of chemical disarmament, 1919–1926. In *Frontline and factory: Comparative perspectives on the chemical industry at war, 1914–1924,* ed. Roy MacLeod and Jeffrey Allan Johnson, 221–45. Archimedes series, vol. 16. Dordrecht, NL: Springer.

Journal of Agriculture and Food Chemistry. 1959. Personal profile: Frank S. Washburn. 7:219.

Kelly, Jack. 2004. *Gunpowder.* New York: Basic Books.

Kiefer, David M. 2001. Capturing nitrogen out of the air. *Today's Chemist* 10 (February):117–22.

Klein, Herbert S. 2003. A concise history of Bolivia. Cambridge, England: Cambridge University Press.

Klemm, Friedrich. 1959. *A history of Western technology.* New York: Charles Scribner's Sons.

Koester, Frank. 1913. *Electricity for the farm and home.* New York: Sturgis & Walton.

Kolber, Zbigniew. 2006. Getting a better picture of the ocean's nitrogen budget. *Science* 312:1479–80.

Ladha, J. K., and P. M. Reddy. 2003. Nitrogen fixation in rice systems: State of knowledge and future prospects. *Plant and Soil* 252:151–67.

Larson, Henrietta M., Evelyn H. Knowlton, and Charles S. Popple. 1971. *New horizons 1927–1950: History of Standard Oil Company (New Jersey)*. New York: Harper & Row.

Lefebure, Victor. 1923. *The riddle of the Rhine*. New York: E. P. Dutton.

Lehrer, Steven. 2000. *Wannsee House and the Holocaust*. Jefferson, NC: McFarland & Co.

Leigh, G. J. 2004a. Haber-Bosch and other industrial processes. In *Catalysts for nitrogen fixation*, ed. Barry E. Smith, R. L. Richards, and W. E. Newton, 33–54. Dordrecht, Germany: Kluwer Academic.

———. 2004b. *The world's greatest fix*. Oxford, England: Oxford University Press.

Lesch, John E., ed. 2000. *The German chemical industry in the twentieth century*. Dordrecht, Germany: Kluwer Academic.

Lochner, Louis P. 1954. *Tycoons and tyrant: German industry from Hitler to Adenauer*. Chicago: Henry Regnery Co.

Lubbock, Basil. 1955. *Coolie ships and oil sailers*. Glasgow, Scotland: Brown, Son & Ferguson.

———. 1966. *The nitrate clippers*. Glasgow, Scotland: Brown, Son & Ferguson.

Lyon, Peter. 1960. The fearless frogman. *American Heritage Magazine* 11. www.americanheritage.com/articles/magazine/ah/1960/3/1960_3_36.shtml.

MacMillan, Margaret Olwen. 2002. *Paris 1919: Six months that changed the world*. New York: Random House.

Markham, Clements R. 1862. *Travels in Peru and India*. London: John Murray.

Markl, Hubert. 2003. Jewish intellectual life and German scientific culture during the Weimar period: The case of the Kaiser Wilhelm Society. *European Review* 11:49–55.

Martin, Geoffrey, and William Barbour. 1915. *Industrial nitrogen compounds and explosives*. New York: D. Appleton and Co.

Martin, Thomas Commerford. 1902. "Fixing nitrogen" from the atmosphere. *American Monthly Review of Reviews* 26:338–42.

Massell, David. 2000. *Amassing power: J. B. Duke and the Saguenay River, 1897–1927*. Montreal, Quebec, Canada: McGill-Queen's University Press.

Masterson, Daniel M. 2004. *The Japanese in Latin America.* Urbana: University of Illinois Press.

Mathew, W. M. 1981. *The house of Gibbs and the Peruvian guano monopoly.* London: Royal Historical Society.

Matthews, Robert. 2000. The Dr. Faust of science. *Focus* (June):114–18.

Maxwell, Gary R. 2004. *Synthetic nitrogen products: A practical guide to the products and processes.* New York: Kluwer Academic.

McArthur, Charles W. 1990. *Operations analysis in the U.S. Army Eighth Air Force in World War II.* Providence, RI: American Mathematical Society.

McConnell, Robert E. 1919. The production of nitrogenous compounds synthetically in the United States and Germany. *Journal of Industrial and Engineering Chemistry* 11:837–41.

McCreery, David J. 2000. *The sweat of their brow: A history of work in Latin America.* New York: M. E. Sharpe.

Meinzer, Lothar. 1998. Productive collateral or economic sense? BASF under French occupation, 1919–1923. In *Determinants in the evolution of the European chemical industry, 1900–1939,* ed. Anthony S. Travis, Harm G. Schröter, Ernst Hamburg, and Peter J. T. Morris, 51–63. Dordrecht, Germany: Kluwer Academic.

Merewether, F. H. S. 1898. *A tour through the famine districts of India.* Philadelphia: J. B. Lippincott.

Mernitz, Kenneth S. 1990. Firms in conflict: Liquid fuel producers in the U.S. and Germany, 1910–1933. *Business and Economic History* 19:143–52.

Miller, Dayton Clarence. 1939. *Sparks, lightning, cosmic rays.* New York: Macmillan.

Miller, Donald L. 2006. *Masters of the air.* New York: Simon & Schuster.

Mittasch, Alwin. 1932. The award of the Nobel Prize in chemistry to Dr. Carl Bosch and Dr. Friedrich Bergius. *The Scientific Monthly* 34:278–83.

———. 1951. *Geschichte der Ammoniaksynthese.* Weinheim, Germany: Verlag Chemie.

Monteon, Michael. 1975. The British in the Atacama Desert. *Journal of Economic History.* 35:117–33.

———. 1979. The enganche in the Chilean nitrate sector, 1880–1930, *Latin American Perspectives* 6:66–79.

Mueller-Hill, Benno. 1988. *Murderous science.* Oxford, England: Oxford University Press.

Myers, Norman, and Jennifer Kent, eds. 2005. *The new atlas of planet management*. Berkeley: University of California Press.

Nagel, Alfred von. 1958. Carl Bosch. In *Ludwigshafener Chemiker,* vol. 1, ed. Kurt Oberdorffer, 109–36. Duesseldorf, Germany: Econ-Verlag.

Nosengo, Micola. 2003. Fertilized to death. *Nature* 425:894–95.

O'Brien, Thomas F. 1982. *The nitrate industry and Chile's crucial transition: 1870–1891.* New York: New York University Press.

———. 1989. "Rich beyond the dreams of avarice": The Guggenheims in Chile. *Business History Review* 63:122–59.

———. 1996. *The revolutionary mission: American enterprise in Latin America, 1900–1945.* Cambridge, England: Cambridge University Press.

Partington, J. R. 1960. *A history of Greek fire and gunpowder.* Cambridge, England: W. Heffer & Sons.

Peck, George. 1854. *Melbourne and the Chincha Islands; with sketches of Lima, and a voyage round the world.* New York: Charles Scribner.

Perutz, Max. 1998. *I wish I'd made you angry earlier.* Cold Spring Harbor, NY: Cold Spring Harbor Laboratory Press.

Principe, Lawrence M. 1995. Newly discovered Boyle documents in the Royal Society archive: Alchemical tracts and his student notebook. *Notes and Records of the Royal Society of London.* 49:57–70.

Reader, W. J. 1970. *Imperial Chemical Industries: A history.* London: Oxford University Press.

Reimann, Guenter. 1942. How Farben swindled Standard Oil. *New Republic* (April 13):483–86.

Roehl, John C. G. 1967. *Germany without Bismarck.* Berkeley: University of California Press.

———. 1994. *The kaiser and his court: Wilhelm II and the government of Germany.* Cambridge, England: Cambridge University Press.

Rotmans, Jan, and Bert de Vries, eds. 1997. *Perpectives on global change: The TARGETS approach.* Cambridge, England: Cambridge University Press.

Saftien, Karl. 1958. Heinrich von Brunck. In *Ludwigshafener Chemiker* vol. 1, ed. Kurt Oberdorffer, 11–30. Duesseldorf, Germany: Econ-Verlag.

Sahota, Gian S. 1968. *Fertilizer in economic development.* New York: Praeger.

Sater, William F. 1973. Chile during the first months of the War of the Pacific. *Journal of Latin American Studies* 5:133–58.

Schmidhuber, Jürgen. (n.d.). Haber & Bosch. www.idsia.ch/~juergen/haberbosch.html

Schneider, Otto. 1909. The oxidation of atmospheric nitrogen. *Journal of Industrial and Engineering Chemistry* 120–21.

Schwendinger, Robert J. 1988. *Ocean of bitter dreams.* Tucson, AZ: Westernlore Press.

Sime, Ruth Lewin. 1996. *Lise Meitner: A life in physics.* Berkeley: University of California Press.

Skaggs, Jimmy M. 1994. *The great guano rush.* New York: St. Martin's Press.

Smil, Vaclav. 1999. China's great famine: 40 years later. *British Medical Journal* 319:1619–21.

————. 2001. *Enriching the earth: Fritz Haber, Carl Bosch, and the transformation of world food production.* Cambridge, MA: MIT Press.

Smith, Barry E., Raymond L. Richards, and William E. Newton, eds. 2004. *Catalysts for nitrogen fixation.* Dordrecht: Kluwer Academic.

Smith, George David. 1988. *From monopoly to competition: The transformation of Alcoa, 1888–1986.* Cambridge, England: Cambridge University Press.

Sobol, Donald J. 1975. *True sea adventures.* Nashville, TN: Thomas Nelson.

Sondhaus, Lawrence. 2004. *Navies in modern world history.* London: Reaktion Books.

Spicka, Mark E. 1999. The devil's chemists on trial: The American prosecution of I. G. Farben at Nuremberg. *The Historian* 865–82.

Stansfield, Alfred. 1914. *The electric furnace: Its construction, operation, and uses.* New York: McGraw-Hill.

Stern, Fritz. 1999. *Einstein's German world.* Princeton, NJ: Princeton University Press.

————. 2006. *Five Germanys I have known.* New York: Farrar, Straus and Giroux.

Stewart, Watt. 1951. *Chinese Bondage in Peru.* Durham, NC: Duke University Press.

————. 1968. *Henry Meiggs, Yankee Pizarro.* New York: AMS Press.

Stocking, George W., and Myron W. Watkins. 1948. *Cartels or competition?* New York: Twentieth Century Fund.

Stokes, Raymond G. 1985. The oil industry in Nazi Germany, 1936–1945. *Business History Review* 59:254–77.

————. 2004. From the IG Farben fusion to the establishment of BASF AG (1925–1952). In *German industry and global enterprise: BASF: The history of a company,* ed. Werner Abelshauser, Wolfgang von Hippel, Jeffrey Allan Johnson, and Raymond G. Stokes, 206–361. Cambridge, England: Cambridge University Press.

Stoltzenberg, Dietrich. 2004. *Fritz Haber: Chemist, Nobel laureate, German, Jew.* Philadelphia: Chemical Heritage Press.

Stranges, Anthony N. 1984. Friedrich Bergius and the rise of the German synthetic fuel industry. *Isis* 75:643–67.

Strathern, Paul. 2001. *Mendeleyev's dream: A quest for the elements.* New York: St. Martin's Press.

Szollosi-Janze, Margit. 2000. Losing the war, but gaining ground: The German chemical industry during World War I. In *The German Chemical Industry in the Twentieth Century,* John E. Lesch, ed., 91–121. Dordrecht, Germany: Kluwer Academic.

Tonitto, C., M. B. David, and L. E. Drinkwater. 2006. Replacing bare fallows with cover crops in fertilizer-intensive cropping systems: A meta-analysis of crop yields and N dynamics. *Agriculture Ecosystems & Environment* 112:58–72.

Travis, Anthony S., Harm G. Schröter, Ernst Homburg, and Peter J. T. Morris, eds. 1998. *Determinants in the evolution of the European chemical industry, 1900–1939.* Dordrecht, Germany: Kluwer Academic.

Travis, Anthony S. 1998. High pressure industrial chemistry: The first steps, 1909–1913, and the impact. In *Determinants in the evolution of the European chemical industry, 1900–1939,* eds. Anthony S. Travis, Harm Schröter, Ernst Homburg, and J. T. Morris, 3–21. Dordrecht, Germany: Kluwer Academic.

Travis, Tony. 1993. The Haber-Bosch process: Exemplar of 20th century chemical industry. *Chemistry and Industry* 15:581–85.

Trumpener, Ulrich. 1975. The road to Ypres: The beginnings of gas warfare in World War I. *Journal of Modern History* 47:460–80.

Tschudi, Johann Jakob von. 1849. *Travels in Peru, during the years 1838–1842.* New York: George P. Putnam.

Tudge, Colin. 2003. *So shall we reap.* London: Allen Lane.

Twain, Mark. 1913. *Roughing it.* New York: Harper and Bros.

U.S. Air Force. 1987. *The United States strategic bombing surveys: European War, Pacific War.* Maxwell, AL: Air University Press.

Van Rooij, Arjan. 2005. Engineering contractors in the chemical industry: The development of ammonia processes, 1910–1940. *History and Technology* 21:345–66.

Waeser, Bruno. 1926. *The atmospheric nitrogen industry,* vol. 1. Philadelphia: P. Blakiston's Son & Co.

Webster, Charles, and Charles Rosenberg eds. 1982. *Joan Baptista Van Helmont: Reformer of science and medicine.* Cambridge, England: Cambridge University Press.

Weintraub, Stanley. 2001. *Whistler: A biography.* New York: Da Capo Press.

West, John B. 2005. Robert Boyle's landmark book of 1660 with the first experiments on rarified air. *Journal of Applied Physiology* 98:31–39.

Willstätter, Richard. 1965. *From my life: The memoirs of Richard Willstätter.* New York: W. A. Benjamin.

Wilmott, Bill. 2004. Chinese contract labor in the Pacific Islands during the nineteenth century. *Journal of Pacific Studies* 27:161–76.

Wisniak, Jaime. 2000. The history of saltpeter production with a bit of pyrotechnics and Lavoisier. *Chemical Educator* 5:205–9.

Wisniak, Jaime, and Ingrid Garces. 2001. The rise and fall of the salitre (sodium nitrate) industry. *Indian Journal of Chemical Technology* 8:427–38.

Yergin, Daniel. 1991. *The prize: The epic quest for oil, money, and power.* New York: Simon & Schuster.

ACKNOWLEDGMENTS

FIRST THANKS GO to my agent, Nat Sobel; Julia Pastore, my editor at Harmony; and my indispensible first editor (and talented writer and wife), Lauren Kessler. Karla Schultz did sensitive work translating technical and biographical works from the German. Thanks as well to the many able archivists and librarians who helped me at the BASF archives, the Carl Bosch Museum, the British Association for the Advancement of Science, the Royal Swedish Academy of Sciences, the Albert Einstein Archives, the University of Oregon, the University of Chicago, Oregon State University (special thanks to Special Collections head Clifford Mead), the Niels Bohr Library and Archives, the Churchill Archives Center, the Bristol Record Office and Central Lending Library, los Museos Naval y Regional de Iquique, and the staffs of numerous smaller historical and anthropological museums in Peru and Chile. I am grateful to the following historians and scientists for their answers and advice: Peter Hayes, Arnold Bauer, Jeffrey Johnson, David Tyler, Anthony Stranges, Eduardo Deves, Thomas P. Hughes, and Ute Deichmann.

INDEX

agriculture, *see* farming
alchemy, 13–15, 79
Allied Chemical, 212, 221, 222
American Cyanamid, 136
ammonia, 66
   price of, 207, 229
ammonia synthesis:
   Haber's work with, 61, 66–69, 75,
     78, 80–81, 88, 89–93, 97–100,
     184–85; *see also* Haber-Bosch
     system
   Nernst's work with, 66, 68–69, 90,
     91, 100, 123
   Ostwald's work with, 78–80, 91,
     94–96, 97, 106, 108, 117, 185
ammonium sulfate, 135, 140, 145,
   201
Atacama Desert, 37–45, 51, 61, 279
atmosphere, xi–xii
   nitrate pollution in, 272, 273,
     275–76
atmospheric nitrogen, xii, xiii, 8,
   66–67
   fixing of, *see* nitrogen-fixing
     processes
Auer, 102, 109
Auschwitz, xvi, 280–81
Austria, 152
automobiles, 208, 224, 228

gasoline for, *see* gasoline
Hitler's enthusiasm for, 242

BASF (Badische Anilin- und Soda-
   Fabrik), 79–88, 89–100, 211
   Bosch as head of, 178, 193–202
   Bosch hired by, 94
   Bosch made head of nitrogen
     research at, 96
   competition faced by, 207
   as defense industry, 140
   as dye company, 82–84, 85–86,
     159, 172, 174
   French and, 172–75, 177–78, 179,
     180, 203, 204–5
   gasoline production and, 209–11
   Haber machine at, 100, 101–9,
     111–23; *see also* Haber-Bosch
     system
   Haber's contracts with, 89–90,
     102–3, 109
   Hoechst lawsuit against, 122–23,
     125, 127–29
   Leuna plant of, *see* Leuna factory
   money printed by, 204
   Nernst's contract with, 128
   Norsk Hydro and, 88, 96, 130
   Oppau plant of, *see* Oppau factory

postwar revival of, 281
profitable period of, 205–6
Schönherr furnaces of, 88, 96, 97, 130
scientific management adopted by, 194–95
sodium nitrate produced by, 139–40, 142–46
Bayer, xv, 87, 211, 212, 230
Beagle, HMS, 19–20, 21, 23
Belgium, 137, 176
Bergius, Friedrich, 209, 210, 211, 213, 221, 232, 242
Bernthsen, August, 92–93, 97, 99, 127, 128
bird guano, 9, 25–36, 38, 40, 53
Bismarck, Otto von, 152, 153
blackbirders, 35
Boer War, 77–78
Bolivia, 43–45
Bosch, Carl, xv–xvi, 93–99, 180–81, 213–14, 241–46, 253–56, 259–64
background of, 93–94
and BASF as defense industry, 140
BASF's hiring of, 94
and British inspectors at Oppau, 173
Brunck and, 103–4
Brüning and, 232–34, 235
collapse suffered by, 200
collections of, 280
conflicts of, 198, 200
death of, 264
Depression and, 229, 230, 232
Deutsches Museum speech of, 262–63
drinking of, 200, 202, 260–61, 263
essay written by, 255–56
as Farben director, 213, 214
France's deal with, 177–78, 179, 180, 206
and French occupiers at Oppau, 172, 173, 174–75, 203, 204–5
gasoline production and, xiv–xv,

208–11, 213, 220–27, 230–32, 234, 243, 244, 255
Haber machine and, 97–99, 101, 102, 104–8, 185; *see also* Haber-Bosch system
at Haber memorial, 258
Haber's correspondence with, 245–46, 250
as head of BASF, 178, 193–202
as head of Farben's managing board, 260
as head of Kaiser Wilhelm Institutes, 261
as head of nitrogen research at BASF, 96
Hitler and, 241, 242, 243–45, 253–56, 258–63
Leuna and, 166, 167, 168, 172, 180, 193–96, 209–10, 218, 219–10, 230–32, 243, 244, 255, 256
Leuna labor revolt and, 193–96
in merger of chemical firms, 211, 212, 213, 219–20, 226
methanol production and, 207
Nazis and, 234
Nobel Prize awarded to, 232
Oppau explosion and, 196–200, 202
Ostwald's machine and, 79, 94–96, 97
political views of, 233, 244
United States trips of, 207–8, 221–22, 223, 224
Versailles peace talks and, 176, 177
villa of, 213, 219, 261
Bosch, Else, 97, 213–14, 261, 263, 264, 281
Britain:
in Boer War, 77–78
chemical and dye industries in, 178–79, 212, 221
fuel production and, 221, 222
Germany and, 151, 173
guano and, 29, 30–31, 32

Haber-Bosch system and, 175–76,
206–7
Haber offered position in, 247–51
India and, 17
nitrate and, 18–19, 39, 40, 52,
54–55, 56, 141
in World War I, 137, 141–42, 158,
163, 168
British Academy of Sciences, 3
British East India Company, 17
Brunck, Heinrich von, 81–88, 93, 95,
96, 98, 103, 113, 193, 194, 195,
221, 225
Bosch and, 103–4, 194, 195
death of, 123, 129
Brüning, Heinrich, 232–34, 235,
241
Brunner Mond, 175, 206, 221, 222
Bücher, Hermann, 244, 262
Bunsen Society for Applied Physical
Chemistry, 65, 69, 75, 80, 123

caliche, 22, 37–38, 39–40, 41, 42, 56
cannons, 15, 16, 114–15, 116
carbon, xii, 278
Charles I, King, 17
chemical weapons, see gas warfare
chemistry, chemical industry, 154,
179, 212, 220
in Germany, 84–85, 104, 125, 154,
161, 171, 179–80, 207–8, 211–13,
243–45
high-pressure, 131, 210, 211, 220
Chile, 20, 35, 42–43
labor demonstrations in, 57–61
nitrate in, *see* sodium nitrate
in nitrate war, 43–50, 51, 53
Peru and, 42–43
World War I battle off coast of,
141–42
China:
coolies from, 27–28, 31
farming in, 5, 6, 269–70
Great Leap Forward in, 269

saltpeter in, 15, 18
starvation in, 269, 270
Chinchas Islands, 25–29, 31, 33, 35,
38
Spanish occupation of, 34–35
chlorine, 84
gas, 159–64, 166
coal, xv, 209, 210, 220, 221, 222, 224,
225
compost, xiii, 5, 6, 7, 8, 9
Council of the Gods, 212, 230
Crimean War, 39
Crookes, Sir William, 3–4, 6–11, 53,
77, 78, 79, 87–88, 100, 108, 269,
271, 277
cyanamid process, 136, 137, 138, 139,
140, 143, 145, 168

Darwin, Charles, 19–23, 36, 38
Davy, Humphrey, 30
Daza, Don Hilarión, 44, 45
Demon Under the Microscope, The
(Hager), xv
Department M, 188–91
Depression, Great, 229–34, 242, 254,
259, 279
Deutsches Museum, 262–63
Diesel, Rudolf, 114
Duisberg, Carl, 211, 212, 230, 260
DuPont, 207, 222
dye industry, 82–84, 85–87, 115, 125,
159, 172, 174, 178–79, 180, 203,
205, 208, 211, 220
indigo, 82–83, 85–86, 103, 225
dynamite, 38

Easter Islands, 35
Ehrlich, Paul, xiv, 165
Einstein, Albert, 72, 132, 154
Bosch and, 214
Haber and, 132–33, 217, 241, 247,
248
marriage of, 132, 133

Nazis and, 237, 239
Elon, Amos, 70
England, *see* Britain
Engler, Carl, 80, 81, 93
Ewald, Paul, 256
explosives, xiv, 38, 39, 51, 77, 87,
  179–80, 185, 200
  in World War I, xiv, 136–39, 141,
  146, 147, 166, 167, 168
  in World War II, xiv, xv, 263, 265,
  268, 281

Farben, *see* IG Farben
farming, xii–xiii, 4–7, 9, 11
  in China, 5, 6, 269–70
  crop rotation in, xiii, 5, 7, 277
  effects of Haber-Bosch on,
  276–77
  fertilizers in, *see* fertilizers
Faust (Goethe), vii, 79
fertilizers, xiv, 6, 7–11
  bird guano, 9, 25–36, 38, 40, 53
  cyanamid, *see* cyanamid process
  manure, xiii, 4–5, 6, 7, 8, 9, 272,
  274, 277
  potassium nitrate (true saltpeter),
  14–17, 18, 39
  sodium nitrate (Chilean nitrate),
  see sodium nitrate
  synthetic, 7–8, 10, 78, 100, 108,
  135; *see also* ammonia synthesis;
  Haber-Bosch system
Fillmore, Millard, 32
fireworks, 15, 23, 38
Ford, Edsel, 223
Ford, Henry, 208
Ford Motor Company, xv, 222, 223,
  224
France:
  dye production in, 178–79
  Germany and, 151, 152
  Germany occupied by, 172–75,
  178, 203, 204
  guano and, 32

Haber-Bosch system and, 172–75,
  177–78, 179, 180, 206, 227
Jews in, 70
nitrate and, 18, 39, 40
in World War I, 137, 141, 143–44,
  158, 160, 162, 166, 169, 176–77
Franck, James, 72, 159, 164, 237–38
fuel, *see* gasoline

gasoline, xiv–xv, 208–11, 213,
  220–27, 229, 230–32, 234, 243,
  244, 255, 260, 261, 263, 265, 266
  German output of, 266
  Hitler and, xv, 242, 268, 281
gas warfare, 158
  Haber's development of, 147,
  157–64, 165, 166, 184, 186–87,
  237
General Motors, 222
Germany, 151–53
  automobile revolution and, 224
  Brüning in, 232–34, 235, 241
  chemical industry in, 84–85, 104,
  125, 154, 161, 171, 179–80,
  207–8, 211–13, 243–45
  civil service edict in, 236, 237, 238,
  244–45, 253, 254
  communists in, 171, 180, 184,
  193–94, 202, 229, 233, 235, 240,
  242, 256
  cyanamid industry in, 137, 138,
  139, 140, 143, 145
  Depression and, 229, 231, 233, 234,
  242, 254, 259
  dye industry in, 82–84, 85–87, 115,
  125, 159, 172, 174, 178–79, 180,
  203, 205, 208, 211, 220
  economic problems of, 203–4, 205,
  227
  French occupation of, 172–75, 178,
  203, 204
  gasoline production in, see gasoline
  Hitler in, *see* Hitler, Adolf
  Jewish judges in, 236

Jews in, 69–71, 72, 73, 126, 155,
 217–18, 234, 236, 237–40,
 244–45, 247, 253, 254
Jews' flight from, 254
merger of chemical firms in,
 211–13, 219–20, 226
Nazis in, *see* Nazis
nitrate needs of, 39, 40, 52–53, 56,
 78, 87, 100, 136–42, 146, 166,
 167, 168
rearmament of, 260
Reichstag in, 154
reparations required of, 177, 187,
 188, 191, 203, 204, 211, 224, 229
revolts in, 171, 184
science in, 73, 154, 185, 238, 240
universities in, 240
Versailles treaty and, 173, 176–77,
 179–80, 184, 187, 240, 254, 260
Weimar Republic in, 171, 187, 203
Wilhelm in, *see* Wilhelm II,
 Kaiser
in World War I, xiv, 133, 136–38,
 141–47, 151, 153–55, 157–69,
 171–72, 176, 236, 240
in World War II, xiv, xv, 264,
 265–68, 281
global warming, xiii, 11, 278
Goethe, Johann Wolfgang von, vii,
 79, 165
gold, 187–91, 215
Göring, Hermann, 242, 262
Grau Seminario, Miguel María, 46,
 47, 48–49, 50
guano, 9, 25–36, 38, 40, 53
Guano Islands Act, 33–34
Guano War, 35, 43
Guinness Brewery, 112
gunpowder, xiv, xv, 15–18, 39, 78, 87,
 100, 137, 138, 140–42, 146, 180,
 185

Haber, Charlotte (née Nathan), 157,
 164, 183, 185–86, 216–17, 218

Haber, Clara (née Immerwahr),
 74–75, 110, 155–57, 160, 163–65,
 251
 suicide of, 164–65, 183
Haber, Fritz, xv–xvi, 65–75, 181,
 183–91, 215–18, 237–41, 246–51
 ammonia work of, 61, 66–69, 75,
 78, 80–81, 88, 89–93, 97–100,
 184–85; *see also* Haber-Bosch
 system
 BASF's contracts with, 89–90,
 102–3, 109
 birth and childhood of, 69, 71
 Bosch's correspondence with,
 245–46, 250
 British position accepted by, 247–51
 conversion to Christianity, 72, 217,
 235, 236, 241, 246
 death of, 251, 253, 257
 Einstein and, 132–33, 217, 241,
 247, 248
 gas weapons developed by, 147,
 157–64, 165, 166, 184, 186–87,
 237
 gold extraction work of, 187–91,
 215
 heart problems of, 217, 218, 235,
 246, 247, 248, 249, 251
 Hitler and, 218, 235–36, 237,
 238–41
 Hoechst lawsuit and, 123, 125,
 127–29
 insecticide research of, 280
 institute of, 109–10, 125–27,
 154–55, 157, 184, 186, 215–18,
 235–36, 237–40, 245, 249–50,
 281
 as Jew, 66, 69–71, 72, 73, 80, 126,
 133, 154, 155, 217–18, 235, 236,
 237, 238, 240, 241, 246–49
 marriages of, 74–75, 155–57, 183,
 185–86, 216–17, 218; *see also*
 Haber, Charlotte; Haber, Clara
 memorial for, 257–59
 money worries of, 218, 235, 250

Nernst and, 65–66, 68–69, 75, 77, 80, 81, 90, 123
Nobel Prize awarded to, 184–85, 241
Oppau explosion and, 200
Ostwald and, 66, 185
resignation of, 239–40, 243, 246, 247, 249
science career pursued by, 73–74
secret negotiations of, 186–87
in Spain, 247
speech made by, 108–9, 125
as university professor, 74, 90
World War I and, xvi, 133, 136–37, 147, 151, 154–55, 157–64, 165–66, 183–84, 246, 250, 251
Haber, Hermann, 74–75, 110, 133, 156, 164, 165, 184, 247, 251, 281
Haber-Bosch system, xiv–xv, 101–9, 111–23, 135–36, 138, 143, 145, 179, 225, 227, 270–78
Britain and, 175–76, 206–7
catalyst in, 105–9, 111–12, 113, 115, 116, 271
in China, 270
France and, 172–75, 177–78, 179, 180, 206, 227
global warming and, xiii, 278
Hitler and, xiv–xv, 261–62
hydrogen problems in, 118–21
Leuna plant for, *see* Leuna factory
nitrate pollution from, xiii, 272–77
Oppau plant for, *see* Oppau factory
prototype machines for, 115–22, 129
raw materials for, 111–13, 116
reactor in, 113–15
sodium nitrate made from, 139–40, 142–46
spread of secrets about, 206–7
today's use of, 271
United States and, 175, 207, 227
Hahn, Otto, 159, 259
Heine, Heinrich, 72

Hertz, Gustav, 159
Hessberger, Johannes, 88
Hindenburg, Paul von, 168, 234, 235, 236
Hitler, Adolf, 218, 229, 234–36, 237, 239–45, 254, 259, 280
appointed chancellor, 257
automobiles and, 242
Bosch and, 241, 242, 243–45, 253–56, 258–63
Farben and, 241–45, 253–55, 259–61, 263, 265, 266, 281
gasoline production and, xv, 242, 268, 281
Haber and, 218, 235–36, 237, 238–41
Haber-Bosch system and, xiv–xv, 261–62
invasions by, 262, 263, 264, 266
Hoechst, 87, 89, 211
lawsuit against BASF by, 122–23, 125, 127–29
Hughes, Thomas Parke, 230
Humboldt, Alexander von, 30
Humboldt Current, 21, 30
hydrogen, xii, 66, 67, 69, 90, 91, 95, 96, 98, 111, 116, 122, 123, 231
Haber-Bosch oven problems from, 118–21
purification of, 112–13

IG Farben, xv
Bosch as director of, 213, 214
Brüning regime and, 233–34
Depression and, 229–30, 231, 234
explosives produced by, 263, 265, 281
Ford and, xv, 222, 223, 224
formation of, 212–13, 214, 219–20, 226
fuels produced by, 220–27, 230–32, 234, 242, 243, 244, 255, 260, 261, 263, 265, 266, 281
Haber memorial and, 258

international deals of, xv, 220–24, 226, 227
Leuna plant of, *see* Leuna factory
Nazis and, 241–45, 253–55, 259–61, 263, 265, 266, 281
Nuremberg trials and, 281
research and development at, 223, 227–28
rubber produced by, 227–28, 260, 265, 281
Standard Oil and, xv, 222–23, 224, 227, 229, 231
Imperial Chemical Industries, 212, 221
India, 17, 18, 19, 39
indigo, 82–83, 85–86, 103, 225
Industrial Revolution, 6, 25, 84
Iquique, 19, 20, 21, 23, 41, 46, 49, 50, 53–54, 56, 61, 279
Iquique Massacre, 58–61
iron, 79, 91, 95, 96, 105, 106–8, 117, 118, 139

Kaiser Wilhelm Institute for Physical Chemistry and Electrochemistry, 109–10, 125–27, 154–55, 157, 184, 186, 215–18, 237–40, 245, 249–50
Haber's resignation from, 239–40, 243, 246, 247, 249
Hitler and, 235–36
renaming of, 281
Kaiser Wilhelm Institutes (KWI), 126, 132–33, 154, 240, 256, 264
Bosch as head of, 261
Haber memorial sponsored by, 257–59
Nazis and, 238, 244, 261
renaming of, 281
Koppel, Leopold, 102, 109, 110, 217, 238
Krauch, Carl, 113, 167, 201, 207, 230, 255, 281
Krupps, 114–15, 116, 118

Laeisz company, 52–53
Lappe, Franz, 101, 105
Laue, Max von, 257
Le Rossignol, Robert, 68, 80, 81, 89–90, 91, 98, 99, 102, 104, 185
Leuna factory, 144–47, 166–68, 172, 175, 177–80, 193–96, 203, 205, 206, 218–20
air raids on, 265–68
current operations at, 281
gasoline production at, 209–11, 220, 221, 222, 224–27, 230–32, 234, 242, 244, 255, 265
Hitler and, 242, 243, 244
labor activism at, 193–96, 202, 218
methanol production at, 207
as money pit, 219–20, 225, 230
shutdown urged for, 230–31
undercover journalist at, 218–19
lightning, xiii, 67, 78, 272, 273
Liverpool Nitrate Co. Ltd., 54

Malthus, Thomas, xiv, 277
Manifesto to the Civilized World, 165–66, 185
manure, xiii, 4–5, 6, 7, 8, 9, 272, 274, 277
bird guano, 9, 25–36, 38, 40, 53
Mao Zedong, 269, 270
Meitner, Lise, 216
methanol, 207
Mittasch, Alwin, 99–100, 101, 105–7, 111, 115, 116, 117, 139, 259, 271
on Bosch, 214
Mond, Sir Alfred, 221, 222
muskets, 15, 16

Nathan, Charlotte, *see* Haber, Charlotte
Nazis, xv, 229, 234–36, 237–45, 253–63, 280
Bosch and, 241, 242, 243–45, 253–56, 258–63

concentration camps of, xvi,
  280–81
Farben and, 241–45, 253–55,
  259–61, 263, 265, 266
Haber memorial and, 257–59
see also Hitler, Adolf
Nernst, Walther, 65–66, 68–69, 75,
  77, 79, 80, 81, 90, 91, 100, 165
  BASF's contract with, 128
  Bosch and, 214
  Einstein and, 132
  Hoechst lawsuit and, 123, 127, 128
Neruda, Pablo, 57
nitrate companies:
  ficha system of, 56–57, 58, 59, 61
  mills of, 56, 61
nitrate pollution, xiii, 272–77
nitrates:
  potassium (true saltpeter), 14–17,
    18, 39
  sodium (Chilean), *see* sodium
    nitrate
nitrate war (War of the Pacific),
  43–50, 51, 53
nitric acid, 38, 78, 137, 138, 139, 140
nitrogen, xii–xiii, 272
  ammonia formation from, *see*
    ammonia synthesis
  atmospheric, xii, xiii, 8, 66–67
  cycles of, 273, 278
  in fertilizers, 8; see also fertilizers
  fixed, xii–xiii, 8–9, 10, 272, 273; *see
    also* nitrates; nitrogen-fixing
    processes
  N_2, 66–68, 99, 118, 121, 136,
    272–73, 275
  purification of, 112
nitrogen-fixing processes, 66–67, 77,
  78, 79–80, 87–88, 220
  ammonia synthesis, *see* ammonia
    synthesis; Haber-Bosch system
  arc, 88, 97, 130
  bacterial, 272, 273
  cyanamid, 136, 137, 138, 139, 140,
    143, 145, 168

nitrogen oxides, 275, 276
nitroglycerine, 38, 137
Nixon, Richard, 270
Nobel, Alfred, 38
Nobel Prize:
  awarded to Bosch, 232
  awarded to Haber, 184–85, 241
Norway, 227
  Norsk Hydro in, 88, 96, 130
North, John Thomas, 53–55
Nuremberg trials, 281

oceans:
  gold in, 187–91, 215
  nitrate pollution in, xiii, 274, 277
oil supplies, 208–9, 225–26
Operation Disinfection, 160–61
Oppau factory, 122, 123, 129–31, 135,
  136, 139–40, 142–46, 177–78,
  206, 223
  air attacks on, 143–44
  explosion at, 196–202
  French occupation and, 172–75, 178
  labor activism at, 202
  sodium nitrate produced at,
    139–40, 142–43, 145, 201
osmium, 91–93, 97, 98, 105, 106, 107,
  109, 116–17
Ostwald, Wilhelm, 66, 77–80, 91, 93,
  94–97, 100, 106, 108, 117, 123,
  127–28, 138, 165, 185
oxygen, xii, 67, 112

Peru, 19, 20
  Bolivia and, 43, 45
  Chile and, 42–43
  guano in, 25–36, 40, 53
  nitrate in, 39, 40–41, 42
  in nitrate war, 43–50, 51
philosopher's stone, 14, 79–80, 100
phosphorus, 8
Planck, Max, 132, 154, 165, 185, 240,
  256–57, 261, 281

Haber memorial arranged by,
257–59
platinum, 138–39
population, xiii–xiv, 4, 6, 11, 271, 276
in China, 270
potassium, 8
potassium nitrate (true saltpeter),
14–17, 18, 39
Prat, Arturo, 46–49

Rathenau, Walter, 70
refrigeration, 112
Riddle of the Rhine, The (Lefebure),
180
rivers and lakes, xiii, 272, 273–75
rubber, 227–28, 260, 265, 281
Rupprecht, Crown Prince of
Bavaria, 161
Rutherford, Ernest, 250

Salisbury, Lord, 152
salitre, see sodium nitrate
saltpeter:
sodium nitrate (Chilean nitrate),
see sodium nitrate
true (potassium nitrate), 14–17, 18,
39
Saltpeter Promise, 140, 142, 146
salts, 14–15
Schmitz, Hermann, 260, 281
Schönherr, Otto, 88
Schönherr furnaces, 88, 96, 97, 130
science, xvi–xvii, 13
chemistry, *see* chemistry, chemical
industry
in Germany, 73, 154, 185, 238, 240
Silva Renard, Roberto, 60, 61
Skaggs, Jimmy, 33
sodium nitrate (Chilean nitrate),
9–10, 11, 17–23, 36, 37–50, 51–61,
78, 87, 88, 100, 108, 130, 135, 279
BASF's production of, 139–40,
142–46, 201

converting to true saltpeter, 38
refining, 39–40
World War I and, 136, 137, 138,
141, 142, 145, 146
Soviet Russia, 186–87, 264
Spain:
Chinchas occupied by, 34–35
gas warfare and, 186
Haber's trip to, 247
Peru and, 43
Spee, Maximilian Graf von, 141–42
Speer, Albert, 266, 267–68
Standard Oil, xv, 208, 211, 222–23,
224, 227, 229, 231
starvation, xiv, 4, 7–8, 10, 11, 100,
271, 277
in China, 269, 270
Stern, Fritz, 165

Tarapacá, 17–23, 25, 27, 36, 39, 40,
41, 45, 51, 54, 58
Teagle, Walter, 223
Time, 232
TNT, xiv, 137, 146
Trotsky, Leon, 187

United States:
anti-German sentiment in, 171, 221
automobiles in, 208
Bosch's trips to, 207–8, 221–22,
223, 224
chemical and dye industries in,
178–79, 212, 221, 222
Farben and, 221
German reparations and, 205
guano and, 30, 31, 32, 33, 36
Haber-Bosch system and, 175, 207,
227
Jews relocated to, 254
nitrate and, 39, 51
stock market crash in, 229
in World War I, 168, 177
uranium, 98, 105, 108

Versailles, Treaty of, 173, 176–77,
179–80, 184, 187, 240, 254, 260
Victoria, Queen, 153

War of the Pacific (nitrate war),
43–50, 51, 53
water, nitrates in, 272, 273–75, 277
Webster, Daniel, 34
Weizmann, Chaim, 246–47, 248–49,
250
wheat, 7, 9, 10, 271, 276, 277
"Where There's a Will, There's Also
a Way" (Bosch), 255–56
Wilhelm II, Kaiser, 70, 73, 137,
151–55, 163, 279–80
abdication of, 169, 171, 193, 279
Willstätter, Richard, 72, 217–18, 235,
236, 241, 247
World War I, 77, 133, 154
battle off coast of Chile in, 141–42
explosives in, xiv, 136–39, 141, 146,
147, 166, 167, 168
France in, 137, 141, 143–44, 158,
160, 162, 166, 169, 176–77

gas warfare in, 147, 157–64, 165,
166
Germany in, xiv, 133, 136–38,
141–47, 151, 153–55, 157–69,
171–72, 176, 236, 240
Haber's role in, xvi, 133, 136–37,
147, 151, 154–55, 157–64,
165–66, 183–84, 246, 250, 251
Oppau plant attacked in, 143–44
Versailles treaty following, 173,
176–77, 179–80, 184, 187, 240,
254, 260
World War II:
Bosch on, 264
explosives in, xiv, xv, 263, 265, 268,
281
Germany in, xiv, xv, 264, 265–68,
281
Guano Islands Act and, 34
Leuna in, 265–68
synthetic fuels in, xv

Zyklon B, 280–81

About the Author

THOMAS HAGER, a science writer, lives in Eugene, Oregon. His previous books include *The Demon Under the Microscope: From Battlefield Hospitals to Nazi Labs, One Doctor's Heroic Search for the World's First Miracle Drug,* and *Force of Nature: The Life of Linus Pauling.*

WITHDRAWN